AN ACRE OF GLASS

J. B. ZIRKER

AN ACRE

A HISTORY AND FORECAST OF THE TELESCOPE

OF GLASS

THE JOHNS HOPKINS UNIVERSITY PRESS

BALTIMORE

© 2005 The Johns Hopkins University Press
All rights reserved. Published 2005
Printed in the United States of America on acid-free paper

2   4   6   8   9   7   5   3   1

The Johns Hopkins University Press
2715 North Charles Street
Baltimore, Maryland 21218-4363
www.press.jhu.edu

Library of Congress Cataloging-in-Publication Data

Zirker, Jack B.
An acre of glass : a history and forecast of the telescope / J.B. Zirker.
p. cm.
Includes bibliographical references and index.
ISBN 0-8018-8234-6 (acid-free paper)
1. Telescopes—History.   I. Title.
QB88.Z57 2006 2005
681'.4123—dc22   2005007786

A catalog record for this book is available from the British Library.

The last printed pages of the book are an extension of this copyright page.

CONTENTS

Color plates appear following page 150

ACKNOWLEDGMENTS

I want to thank Trevor Lipscombe, editor-in-chief at the Johns Hopkins University Press, for his helpful criticism and support throughout the writing of this book, our fourth collaboration. The text was the easy part, as it turned out; the illustrations were the devils. Tara Lenington and Juliana McCarthy on the JHUP staff gave me their patient, professional help in finding suitable images and obtaining the necessary permissions to publish them. I thank them heartily. Mary Yates, copy editor, went over the manuscript with an eagle's eye. The book reads much better for her efforts.

Finally, I owe a debt of gratitude to all the astronomers out there who cheerfully supplied information and images. May they prosper.

AN ACRE OF GLASS

PROLOGUE

I AM STANDING ON Mauna Kea, 4,200 meters above the sea, watching the Hawaiian twilight fade into evening. The first stars are beginning to appear in the deep blue vault above me. Far below, a fleecy cloud deck covers the island. A slight breeze is picking up. The night will be clear and moonless, and it promises good hunting for the astronomers.

The top of this extinct volcano is studded with giant telescopes. Behind me the two domes of the Keck telescopes are opening. Inside rest the largest mirrors in the world, 10 meters across, twice the diameter of the famous 200-inch mirror on Palomar Mountain, California. Tonight this triumph of human engineering will peer into the farthest reaches of space.

Somewhere overhead the Hubble Space Telescope is completing another orbit around the Earth. By today's standards the Hubble is a small telescope, a mere 2.5 meters in diameter. It has the incomparable advantage, however, of cruising above the Earth's turbulent atmosphere, capturing images that are nearly as sharp as physics allows. I can picture the team of astronomers grouped around the Hubble's controls at the Space Telescope Science Institute in Baltimore, Maryland. Tonight they will use this superb instrument to examine the black hole at the center of a distant galaxy.

Large telescopes evoke a unique response in most people. They symbolize pure science, our urge to understand our world, in a way that nothing else does. They are tools that expand not just our visual horizons but our mental horizons as well. In a sense they are the grandest monuments of our technical civilization. And they are physically awesome—huge, massive, yet so well balanced that they move at the touch of a finger. Their jewel-like mirrors are among the most finely crafted objects we possess.

The telescopes on this mountain and others like it around the world employ the best of today's technology. But what of tomorrow? What will the next generation of telescopes look like? What will they enable us to see, to find, to understand?

Throughout history astronomers have wanted to look ever deeper into the dark abyss of space, to see ever fainter objects. They have struggled to build larger telescopes, to gather more and more of the faint light these distant objects emit. As their instruments grew in size, astronomers gained a secondary benefit: they could distinguish smaller and more intricate details. Galaxies viewed through early telescopes appeared as amorphous blobs of light, but with larger telescopes these images resolved into chains of stars and clouds of dust that earlier astronomers could only have imagined.

The first need, to see faint objects, has always competed with the second need, to resolve fine details. These parallel themes persist throughout the history of the telescope, and at all wavelengths, from visible to radio. Radio astronomers have built huge dishes to meet the first need and huge interferometric arrays of smaller dishes to meet the second. Now the optical astronomers are following the same route. As we shall see, the two paths have merged, supplementing each other to the benefit of astronomy as a whole.

The Keck telescopes are the now largest in the world, but they won't be for long. Giant mirrors 30 meters, 50 meters, and even 100 meters across are being planned. Meanwhile, the National Aeronautics and Space Administration is planning to put the James Webb Space Telescope, 6.5 meters in diameter, into orbit between the Earth and the Sun. And sometime in the future NASA may launch a huge optical interferometer into space to search for Earth-like planets.

These are exciting times for telescope designers and for the astronomers who depend upon them. In the pages that follow we'll meet some of these people, find out why they want ever-larger telescopes, and learn how these glorious instruments will be built.

We'll begin in chapters 1 and 2 with a glance backwards, tracing briefly the first three hundred years of telescope construction and the discoveries those early designs made possible. In chapter 3 we'll describe some of the key discoveries made at x-ray, radio, and infrared wavelengths between 1950 and 1980, and the questions these discoveries raised for future research. Then in chapters 4 and 5 we'll backtrack to pick up the story of the founding of the great astronomy research centers and the construction of the Hubble Space Telescope. The revolutions in mirror making will be told in chapters 6 and 7, and their consequences in the chapters that follow.

THE FIRST THREE HUNDRED YEARS

TYCHO BRAHE (1546–1601), the great Danish astronomer, died a mere seven years before the first telescope was built. Never dreaming of the wonders that the small, toylike gadget would reveal, he had to rely on the tried-and-true instruments that astronomers and astrologers had used throughout the Middle Ages (fig. 1.1): the *sextant,* to measure the angle above the horizon of an object; the *azimuthal quadrant,* to measure both vertical and horizontal angles; and the *armillary sphere,* to measure angles with respect to the ecliptic and the meridian (note 1.1). Making observations with such devices was a painstaking business. To use the sextant, for example, one had to squint along a radial bar through a sort of gunsight and then measure the bar's angular position on a graduated brass arc. Sometimes two observers were needed, one to squint and one to measure.

Tycho built his own instruments, refining them to a high level of performance. He could take angular measurements as precise as half a minute of arc, more exact than anyone had ever achieved. Every clear night he and his assistants would be out on the roof of his private observatory, Uraniborg, observing the Moon and the planets with respect to the distant stars. Like observers before him, he was puzzled by the apparent reversal of direction of Mars, Jupiter, and Saturn (note 1.2), but he persisted, tracking their motions year after year.

Tycho was wedded to Ptolemy's *geocentric* (Earth-centered) model of the planetary system, but to account for the planets' retrograde motion, he developed his own version of the model. His system had the planets and a small

Fig. 1.1. Some of Tycho Brahe's instruments: *(top left)* sextant; *(top right)* quadrant; *(bottom left)* equatorial armillary circle; *(bottom right)* armillary sphere

comet revolving about the Sun, and the Sun (and the Moon) revolving about the Earth. One more step and Tycho might have arrived at Copernicus's *heliocentric* (Sun-centered) model, but he never went that far.

Shortly before his death Tycho told his young assistant, Johannes Kepler (1571–1630), to calculate an orbit for Mars, whose retrograde motion was especially large. Kepler, unlike Tycho, was a devoted Copernican and approached his task with a heliocentric model in mind. Tycho's data were so precise that despite his lack of optical equipment Kepler was able to make several fundamental discoveries (note 1.3). The simplest, and the one most devastating to the Ptolemaic model of a geocentric universe, was that the orbits of the planets are ellipses, with the Sun at one focus. Extracting these empirical laws from the data was no mean feat. Kepler was able to publish his results only long after Tycho's death.

Tycho Brahe's work represents the high point of pre-telescopic astronomy. From our twenty-first-century perspective it is all too easy to overlook the important contributions of such dedicated workers who struggled to break with their preconceptions and to match their ideas to the data. If we look even further back, some of the insights gained by the ancients are truly astonishing, given their technology. The Chaldeans, for example, discovered the eighteen-year cycle of total solar eclipses nearly three thousand years ago, the Greek Aristarchus (third century B.C.) made the first estimates of the size and distance of the Sun, and another Greek, Hipparchus (second century B.C.), discovered the precession of the equinoxes (note 1.4).

THE DAWN OF A NEW AGE

The telescope developed from magnifying glasses and spectacles. Venice has long been famous for its glasswork. By the mid-fourteenth century the quality of Venetian glass was good enough to see through easily, free of discoloration and bubbles. By 1350, Venetian opticians were grinding convex lenses, mounting them in frames, and selling them to nearsighted scholars. By 1450, opticians had learned to grind concave lenses. At that point it would have been an easy step to combine two lenses, of any type, to create a telescope, but if anyone actually did that, the news was never published. It took another 150 years for the invention to surface.

Nobody knows for certain who invented the first telescope. Most historians seem to favor Hans Lippershey (ca. 1570–1619), a Dutch maker of spectacles. In 1608 he mounted a concave and a convex lens in a tube, and presto! He had a device that magnified a distant object four times. (To see how a telescope works, see figure 1.2 and note 1.5.) Both Lippershey and Jacob Metius, another independent thinker, applied for patents to the Dutch government. The patent board rejected their applications, saying that the gadgets were too simple to patent. As a consolation prize, however, Lippershey received a commission to build several binocular versions.

Lippershey's design spread rapidly throughout Europe, and soon opticians in Paris and Italy were selling copies of his prototype. One version we would recognize as a mariner's spyglass. Thomas Harriot, the scientific aide to Sir Walter Raleigh on his voyage to Virginia, built a 6-power version of the device and used it to study the Moon. His drawings of the Moon, dated August 1609, were the first ever published.

But it was Galileo Galilei (1564–1642), the Italian scientist *extraordinaire,* who proved the maxim that "in the country of the blind, the one-eyed man is king." Galileo's father had wanted him to study medicine, but the young man preferred mathematics. He did so well in this subject that after graduation he was appointed a professor at a succession of universities, at higher and higher salaries. His career took a sharp turn after eighteen years of teaching mathematics at the University of Padua, when he heard of Lippershey's invention.

Like Lippershey, Galileo was a skilled lens grinder and was able to build his

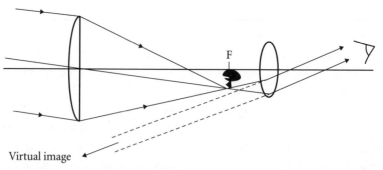

Virtual image

Fig. 1.2. A basic refracting telescope employs a long-focus lens (the objective) and an eyepiece. The objective lens forms a real image of the target at its focal point (F), which coincides with the focal point of the eyepiece. The image seen by the eye is magnified by a factor equal to the ratio of the focal lengths and appears to float at infinity.

own telescope, with a *plano-convex objective* and a *plano-concave eyepiece* (in other words, the lenses were curved on one side and flat on the other). The ratio of the focal lengths of the lenses determined the telescope's magnification. Galileo built a 3-power spyglass in 1609 and later an 8- and then a 20-power version. A magnification of 20 revealed fine details, but the corresponding angular field of view was only 15 arcminutes, or half the width of the full Moon. (An *arcminute* is an angle equal to 1/60 of a degree.) Nevertheless, Galileo made history with this 20-power telescope. Among his discoveries were Saturn's rings (but see note 1.6), the Moon-like phases of Venus, and sunspots. He also learned that the Milky Way is composed of stars and that the Moon has mountains.

Galileo's discovery of the four brightest satellites of Jupiter in November 1609 was perhaps his most important, for it changed the existing view of the universe. Here was an example of objects that orbited about a parent body other than the Earth, a phenomenon that definitely contradicted Ptolemy's geocentric planetary model and supported the heliocentric model of Copernicus. Kepler, who had become a luminary in his own right, was able to confirm this result independently.

The discovery was a sensation and lofted Galileo's career. He described his discovery in his book of 1611, *Siderius Nuncius (The Starry Messenger)*, which he dedicated to the grand duke of Tuscany, Cosimo II de' Medici. The flattery worked, and Galileo was appointed philosopher and mathematician to the grand duke. Fame and fortune followed him everywhere.

Galileo was careful to avoid endorsing the Copernican model as anything more than a convenient mathematical device until 1632, when he published his *Dialogue Concerning the Two Chief Systems of the World—Ptolemaic and Copernican*. But he recanted his views when brought before the Inquisition and remained under house arrest for the rest of his life.

Several scientists throughout Europe attempted to confirm Galileo's discoveries, but they lacked sufficiently good equipment. Christoph Scheiner (1575?–1650) was an exception. This German professor of mathematics was also handy as a builder of instruments. He constructed a telescope consisting of a large convex lens (the *objective*) and a convex eyepiece. This design produced an inverted image, but with a larger field of view than Galileo's spyglass.

Scheiner confirmed Galileo's discovery of sunspots around 1611, but as a Jesuit he subscribed to Aristotle's view that the heavenly bodies (excluding the blotchy Moon) are perfect in all respects. He therefore proposed that sunspots were small planets orbiting close to the Sun. Galileo learned of Scheiner's ideas through a common acquaintance and countered with several arguments. He pointed out that some sunspots appear, change shape, and disappear during their passage across the solar disk. That suggested that they lie near the Sun's surface and are not little planets. Scheiner lost the debate, but his telescope design became known as the astronomical telescope and was preferred to Galileo's optical arrangement.

Scheiner later changed his mind and accepted that sunspots lie on the Sun's surface. In fact, he continued to observe sunspots for fifteen years and discovered that the Sun's axis of rotation is not perpendicular to the ecliptic, the plane of the Earth's orbit, but is instead tilted by an angle of 7.25 degrees.

Galileo is known principally for his astronomical discoveries, but his experiments with pendulums, with balls rolling down inclined planes, and (anecdotally) with objects falling from the Leaning Tower of Pisa paved the way for Sir Isaac Newton's great insights into the laws of mechanics. This Italian genius was also a practical man who invented the proportional compass (a survey tool) and a machine to raise water levels.

HUYGENS

After Galileo's successes, many other scientists began building telescopes. Christian Huygens (1629–95), a native of Holland and another all-around genius, was among the first. Huygens began his career as a mathematician (he founded the subject of mathematical probability). He also proposed a wave theory of light and showed how it explained such properties as reflection, refraction, and diffraction (note 1.7). In 1689 he traveled to London to confer with Isaac Newton, whose work he greatly admired. The two brilliant men debated politely on the nature of light, Newton being convinced that light consists of tiny *corpuscles,* or particles, Huygens upholding his wave theory. In the twentieth century we learned that both men were correct.

Huygens's interest in light prompted him to grind his own lenses and build a series of telescopes. To advance beyond Galileo's discoveries, he needed higher magnification. And to achieve that, he could grind either eyepieces with shorter focal lengths or objectives with longer focal lengths. He chose the easier alternative and made long objectives.

Huygens's grandest achievement was a 123-foot telescope. With this giant he discovered Titan, a satellite of Saturn, and observed the Orion Nebula. As a sideline, Huygens built and obtained a patent for the pendulum clock. He also invented an improved eyepiece for a telescope. It consists of two plano-convex lenses of the same kind of glass, which focus light of all colors to nearly the same point.

In the 1670s the Polish astronomer Johannes Hevelius (1611–87) went to ridiculous extremes in the quest for high magnification. He had established his reputation mapping features on the Moon with a 10-foot telescope. When he heard of Huygens's big guns, he built a telescope 60 feet long, then a giant 140-footer. It was hung on a tower and hauled about with cables by assistants, who struggled against the wind to keep it pointed. Hevelius may have discovered the Andromeda Nebula, but not with this monster.

REFLECTORS

All telescopes built up to this time were *refractors,* which depended on the power of glass to bend light rays. James Gregory (1638–75), a young Scottish mathematician at Saint Andrew's University, conceived the first *reflecting* telescope, which depended on the power of a mirror to reflect a light ray. Gregory's design of 1661 (fig. 1.3, bottom) was far ahead of its time. Lacking the necessary skills, Gregory never built one.

In Gregory's instrument the main mirror has a parabolic shape, ideal for bringing parallel rays from a distant source to a pinpoint focus (note 1.7). A concave secondary mirror, ground to an elliptical shape, is located beyond the focal point of the primary. The secondary reflects the beam through a hole in the primary and finally to an eyepiece. This convenient design halves the length of an equivalent refracting telescope and has the added advantage of focusing all colors to the same point. It is, in short, *achromatic.*

Isaac Newton (1642–1727), perhaps the greatest scientist of his century, was the first to prove that "white" light is composed of a spectrum of colors, the familiar rainbow. In a famous experiment he passed a thin ray of sunlight through a prism and saw the spectrum of colors. To demonstrate that the colors were not generated in the prism, he inserted a second prism, allowed the spectrum to pass through it, and saw that the colors folded back into white light again.

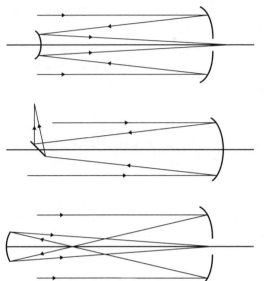

Fig. 1.3. Three types of reflecting telescope: *(top)* Cassegrain; *(middle)* Newtonian; *(bottom)* Gregorian. Each of these telescopes has a parabolic primary mirror and a small secondary mirror to bring the light to a focus through a hole in the primary. In the Cassegrain the secondary is a convex hyperboloid, located between the primary and the prime focus. In the Gregorian the secondary is a concave ellipsoid, located outside the prime focus. The Gregorian design allows easier access to the prime focus, which is useful for testing the entire optical system.

Newton experimented with refracting telescopes but found them unsatisfactory because they brought each color to different focal point, a problem known as *chromatic aberration*. The reason, Newton learned, is that glass bends blue light more than red light, so that a single lens, used as an objective for a telescope, forms a blue image in front of a red image of the same object. The result is a confusing blur. The problem is not so severe in long-focus telescopes, which may have been one of their few advantages.

Dissatisfied with refractors, Newton built the first reflecting telescope in 1668, the year after his monumental theory of gravitation was published in his *Principia*. Unlike Gregory's design, Newton's primary mirror had a spherical surface, which is easier to grind than a parabola, although it introduces other aberrations. Instead of glass, Newton used speculum metal, a shiny alloy of copper and tin (named for *speculum*, the Latin word for "mirror"). Unlike Gregory, he inserted a flat diagonal mirror to deflect the image from the primary to the eyepiece near the front of the telescope (fig. 1.3, middle).

Newton's mirror was only 1.5 inches in diameter, hardly big enough for making new discoveries. Surprisingly for a man of his broad interests, he never built larger telescopes or made observations with them. He was content, instead, to draw upon Tycho Brahe's visual observations and Kepler's empirical laws to formulate his dynamical theory of the motions of the planets.

In 1672 a talented French mathematician, Guillaume Cassegrain, invented yet another reflecting telescope design, which today bears his name (fig. 1.3, top). It has a concave parabolic primary mirror and a convex hyperbolic secondary. Like the Gregorian design, it was very compact and has remained popular to this day.

Although Newton, Gregory, and Cassegrain each knew that the ideal shape for an astronomical mirror is parabolic, none of them was capable of grinding one. It remained for James Short (1710–68), a young Scot, to produce parabolic and elliptical mirrors of high quality. Short was studying for a career as a minister at the University of Edinburgh in the 1720s when he heard the lectures of Colin Maclaurin, the famous mathematician, and was inspired to switch to a career in optics. Through hard work and much experimentation he taught himself to polish aspheric speculum mirrors, primarily for Gregorian telescopes. Eventually he established a business in London, where he produced more than thirteen hundred telescopes over a period of thirty-five years. His sold his fine work to rich patrons throughout Europe, Russia, and Turkey. His largest telescope, with a mirror 24 inches in diameter, was built for the king of Spain.

By Short's time, major observatories had been established in cities such as Paris (1667), Greenwich (1675), and Berlin (1711). Astronomers were slow to adopt the reflecting telescope, however, because speculum metal tarnishes easily and requires constant attention. In fact, the refractor held its place well into the nineteenth century. For example, the Dorpat Observatory (Estonia) installed a 9.5-inch refractor in 1817, and the Pulkovo Observatory (St. Petersburg) acquired a 15-inch refractor, at that time the largest in the world, in 1839.

HERSCHEL'S GIANTS

The reflecting telescope came into its own only gradually. William Herschel (1738–1822) was one of its main proponents. Herschel was born in Germany but lived most of his life in England. Although he was a professional musician who earned his living as an organist and conductor, his real passion was astronomy. He taught himself to grind mirrors (see note 1.8) and was soon making excellent parabolas. In 1781 this talented amateur discovered the planet

Uranus with his little 6-inch mirror, thereby doubling the size of the known solar system. This feat earned him a knighthood, the position of private astronomer to King George III, and a pension that enabled him to devote himself full time to astronomy.

Around this time a friend of Herschel's gave him the gift of a catalog of nebulous objects in the sky that had been compiled by the French astronomer Charles Messier. Messier was a comet hunter, and he cataloged nebulae primarily to avoid confusing them with new comets. His discoveries of comets are largely forgotten, but his list of one hundred nebulae has remained one of the tools of astronomers to this day. The Andromeda Nebula, for example, is often referred to as M31, the *M* standing for Messier.

Herschel was intrigued by Messier's catalog and decided to search for more of these fuzzy objects. He continued to build bigger and bigger reflecting telescopes, achieving a 24-inch and finally a 48-inch Goliath. Such large mirrors couldn't be cast in glass with the technology of the day, so Herschel cast them in the traditional speculum metal. The 24- and 48-inch telescopes were mounted in triangular trusses (fig. 1.4) that rotated about a vertical axis, and the tubes were raised and lowered with a system of cables.

Fig. 1.4. Herschel's 40-foot-long 48-inch telescope

These were the first *altitude-azimuth* mountings, whose principle is still applied to the large telescopes being built today. The arrangement has the disadvantage of requiring nonuniform rotations around both axes to follow an object in the sky. With present-day computers this is no problem, but in Herschel's day it posed a challenge. One early solution to the problem was the development of an alternative mounting, the *equatorial*, the first of which was built in Dorpat in 1824. The equatorial has one axis (the *polar*) parallel to the Earth's axis of rotation and the other (the *declination*) perpendicular to it. Once the telescope is pointed toward an object, the astronomer need only turn it about one axis, the one parallel to the Earth's.

To avoid the need for a secondary mirror, Herschel tipped his primary mirror at an angle to shoot the image off to the side of the telescope tube. He would then position himself near the front end of the tube, some 20 feet off the ground, to look through the eyepiece.

Herschel's large telescopes were too cumbersome to follow the motion of a star across the sky, so he pointed them due south at a fixed altitude above the horizon and noted the time at which a star crossed his field of view. In this way he could scan a complete east-west strip of sky, night after night. He called out his observations to his sister Caroline, who noted them down carefully. This sister-and-brother team managed to map the whole sky visible from England with this method.

Among Herschel's many discoveries is a collection of twenty-five hundred "nebulous" (non-star-like) objects, which are listed in his *Catalog of Star Clusters and Nebulae.* He believed that these fuzzy objects were clusters of many stars and speculated (following Immanuel Kant) that they were "island universes." Many were indeed shown later to be galaxies. Herschel's greatest contribution, however, was his model of the Milky Way, which he derived from his patient counts of stars in different directions. He concluded that our galaxy has the shape of a disk, with a thickness about one-sixth of its diameter.

Nebulae were not Herschel's only interest. He discovered over a thousand double stars and was the first to recognize that these stars orbit each other. He also detected the Sun's motion relative to the stars. He found that the Sun is moving toward a point in the constellation Hercules not far from the bright star Vega. In addition, he discovered the sixth and seventh satellites of Saturn and the first two of Uranus.

Herschel's 48-inch telescope turned out to be too unwieldy to operate, and he soon returned to using his favorite 24-inch. The 48-inch held its place as the largest anywhere, however, until 1842. In that year a 72-inch mirror was cast by William Parsons, the third earl of Rosse, master of Birr Castle in central Ireland.

THE GREAT REFLECTOR

William Parsons (1800–1867) graduated in 1822 from Magdalen College, Oxford, with first-class honors in mathematics. After a brief career in Parliament he decided to devote himself to his real passion, astronomy. Inspired by Herschel's success, he decided to build his own large telescope at Birr Castle, and rather than employ a team of experts, he would cast his own mirror. It would be 36 inches in diameter. At first Parsons tried to assemble a mirror blank from wedge-shaped pieces of speculum metal soldered to a sheet of brass. He gave up this approach after several disasters and eventually cast a single 36-inch blank.

The usual method of grinding a telescope mirror by hand is to rub two blanks against each other, with an abrasive compound between them. As one blank becomes concave (the mirror), the other becomes convex. This is a long, tedious process for even a small mirror. Parsons decided that his blank was too large to shape into a parabola by hand, so he invented a steam-powered grinding machine. The mirror blank rotated slowly as the polishing tool was rubbed back and forth across the blank by a mechanical arm. After many trials and delays, the machine finally produced an acceptable mirror in 1839. It was mounted at Birr Castle and was given high marks by several well-known Irish astronomers.

Parsons used his telescope for only a short time before embarking on a larger one. Perhaps his most important contribution with the 36-inch, aside from a more detailed map of the Moon, was his discovery that he could almost resolve stars in the Andromeda Nebula, something even Herschel had not been able to do. Was the Andromeda a different kind of object than other, featureless nebulae? A larger telescope would be needed to settle the matter.

Let me explain. The *resolving power* of a telescope is its ability to separate very close objects in the sky, like binary stars. Resolution increases as the diameter of the telescope is increased, and also as the wavelength of light decreases. A big primary mirror viewing an object in blue light would be ideal.

Unfortunately, turbulence in the Earth's atmosphere, not the diameter of the mirror, is usually the limiting factor in resolution. For example, Parsons's 36-inch objective, if perfectly *figured* (ground to the required parabolic shape), could resolve two stars only 0.16 arcsecond apart. But the blur produced by the atmosphere might very well limit the resolution to several arcseconds. The great advantage of a big mirror is its ability to gather more light and see fainter objects, not necessarily to resolve two close objects. (An *arcsecond* is an angle equal to 1/60 of an arcminute, or 1/3600 of a degree. Jupiter's diameter as viewed from Earth is about 45 arcseconds, and the headlights of a car 200 miles away are about 1 arcsecond apart.)

Parsons was more interested in building large telescopes than in using them himself. He conceived a plan for building a first-rate telescope and making it available to professional astronomers. It would use a 72-inch mirror, which would handily surpass Herschel's 48-inch and might lead to even greater discoveries.

Parsons built three crucibles, each 24 feet across, to melt the tons of speculum metal he required. The crucibles were placed in furnaces and heated with burning turf. A 4-ton blank was cast successfully in 1842. After a month on the grinding machine, however, the blank cracked. Undaunted, Parsons melted the blank down and recast it, and this time it survived grinding and polishing.

The new reflector, called the Leviathan of Parsonstown, had a focal length of 58 feet, much too large for even a Herschel mount. So Parsons (now the earl of Rosse) built two massive walls, each 50 feet high, and suspended the 8-foot-diameter tube between them with cables. The telescope could only point to a narrow vertical strip of sky in the south (the *meridian*), but the rotation of the Earth would allow Parsons to view all the stars in the sky, as Herschel had done.

The optical quality of the great reflector passed all expectations. It resolved stars at the edges of the Andromeda Nebula, confirming Parsons's earlier hunch. The telescope's greatest discovery, however, was the spiral structure of some twenty nebulae. The first spiral observed was Messier 51, the beautiful Whirlpool Galaxy.

The Leviathan of Parsonstown remained the largest telescope in the world until 1917, when, as we shall see, the 100-inch was commissioned at Mount Wilson Observatory in California. Parsons eventually donated the Leviathan to the Armagh Observatory, Ireland's principal astronomical center.

Parsons had a role in the construction of another large telescope. In 1852 Thomas Romney Robinson, director of the Armagh Observatory, was asked to chair a committee of the Royal Society of London, to advise on the construction of a large reflector for the observatory at Melbourne, Australia. Robinson was a close friend of Parsons's neighbor Thomas Grubb, an optician whose prestigious Dublin firm had built several small telescopes for Armagh. Robinson chose Parsons and the astronomer Walter de la Rue for his committee.

Robinson's committee recommended building a Cassegrain telescope with a 48-inch speculum mirror. Grubb received the contract, ultimately worth about £13,000, and his son Howard supervised the construction of the telescope (fig. 1.5). The design incorporated an equatorial mount, vastly superior to Parsons's alt-azimuth. The telescope was completed in 1868 and was judged magnificent by the Robinson committee. It was used for several decades to study nebulae in the southern sky but proved difficult to operate in any but the lightest winds. The firm of Grubb-Parsons continued to make all the largest British telescope mirrors until the late 1980s.

Fig. 1.5. The Great Melbourne Reflector, with a 48-inch mirror of speculum metal

THE AGE OF THE GREAT REFRACTORS

One might think that the successes of Herschel and Parsons would have persuaded astronomers to shift their loyalties from refracting to reflecting telescopes. A large mirror was plainly the most direct way to collect more light,
and it was achromatic into the bargain. But the huge reflectors were too large
to point anywhere but south, and they required the astronomer to work at a
dangerous height above the ground. Conservative astronomers therefore remained wedded to the refractor. They were aided in this choice by the invention of the *achromatic objective lens* (the primary, or largest lens) nearly a hundred years earlier.

Chester Moore Hall (1703–71), an English lawyer, was an unlikely person to
discover the principle of the achromatic lens. He had learned that a new type
of glass, "flint," refracted light more strongly than the more common "crown"
glass. He knew that while a convex lens brings light to a focus, a concave lens
spreads it out in a cone. Hall realized that a flint concave lens placed behind a
crown concave lens might cancel the chromatic aberration of the crown and
bring rays of several colors to a common focus. He had discovered the principle of the achromatic lens, and in 1729 he decided to have one built.

To keep his valuable invention secret, Hall contracted with two different
lens grinders to fashion the lenses separately. Unfortunately, they each gave the
task to the same subcontractor, who stole the idea. Hall nevertheless got his
lenses and built the first truly achromatic refractor. He never bothered to
patent the idea, however. Some years later John Dolland, another optician,
learned about Hall's achromat, built his own, and took out a patent on the design. Dolland became a rich man, leaving Hall with the credit but none of the
cash.

From this year, 1730, achromatic telescopes were in increasing demand, despite their remaining imperfections. They were difficult to make, however, in
any size larger than a few inches.

Joseph Fraunhofer (1787–1826) was one of the few opticians capable of this
exacting work. Fraunhofer had raised himself from poverty to become the co-
owner of a prestigious optics firm, comparable to the Carl Zeiss of our time.
He was a meticulous craftsman who studied every aspect of lens making in order to improve the product. He invented special grinding and polishing ma-

chines and various tools for measuring the curvature of a lens. He varied the composition of the glasses and studied their refractive properties. In 1824 he built his finest achromatic telescope, 14 feet long and with a 9.5-inch lens, for the observatory at Dorpat. Friedrich Struve (1793–1864), one of the giants of nineteenth-century astronomy, used this superb instrument to measure the separations of over three thousand binary stars.

In addition to refining the craft of lens making, Fraunhofer also carried out a mathematical investigation of the diffraction of light and invented a device for spreading light into its constituent colors, the *diffraction grating*. This device consists of a series of closely spaced parallel lines engraved in glass. Fraunhofer was eventually able to rule as many as four hundred lines per millimeter across a glass several centimeters wide. With this powerful tool he discovered and mapped the dark absorption lines in the solar spectrum, which bear his name (see note 1.9 on Kirchhoff and spectroscopy). King Ludwig of Bavaria knighted Fraunhofer in 1824 for his outstanding discoveries.

The refracting telescope advanced still further with the work of Alvan Clark (1804–87), a Boston portrait painter. Clark had two sons, George Bassett and Alvan Graham, and two daughters, Maria Louisa and Caroline Amelia. Son George became interested in telescopes, and father Alvan followed suit. He soon decided to forgo painting and devote himself instead to astronomical optics. Over several years of hard work he taught himself to grind lenses. He recognized that they were of excellent quality and tried to sell them, but without success. So he and his sons used them to build several small telescopes for themselves, mainly as a hobby.

Clark's lenses turned out to be so good that they were able to resolve double stars closer than any seen before. When Clark published his results, he attracted the attention of the English astronomer William Dawes, who invited him to London. There he met William Parsons and William Herschel's eminent son, John. After that, his future was secure. He returned to Boston and in 1850 established the firm of Clark and Sons. Father and sons went on to craft dozens of superb achromatic objectives, ranging upward from 6 inches. Their work was in great demand. The firm was contracted to build an 18.5-inch for the Dearborn Observatory of Northwestern University, in Evanston, Illinois. Completed in 1862, the Dearborn telescope was the largest refractor in the world.

Alvan Clark senior had the additional satisfaction of making a discovery with this telescope. While testing the new telescope, he saw that Sirius, the brightest star in the northern sky, had a dim companion at a distance of 11 arcseconds. Sirius was in fact a double star, a finding that Friedrich Bessel had predicted in 1841. Bessel, a German astronomer and mathematician, had detected periodic motions of Sirius that implied the existence of a dark companion star. Sirius B, as it is called today, was eventually identified as the first white dwarf (note 1.10).

The U.S. Naval Observatory, in Washington, D.C., was the Clarks' next customer for an outstanding telescope: a 26-inch achromat that in 1873 took its place as the world's largest refractor. Asaph Hall, one of the observatory's astronomers, discovered the two moons of Mars with this telescope in August 1877, when the planet was at its nearest approach to the Earth. He named them Phobos (Fear) and Deimos (Panic), suitable companions to the god of war.

At this same approach of Mars, the sharp-eyed Italian astronomer Giovanni Schiaparelli (1835–1910) drew some excellent maps of Mars's surface markings. He gave fanciful names to the new features he saw, such as Mare Sirenum (Sea of Sirens), Elysium, and Arabia. Among his discoveries were some linear features that he called *canali* (channels). The English translators rendered the word as *canals,* which suggested the transfer of water. Soon a mythology was created. These canals had been built by an advanced civilization to transport Mars's sparse water. Schiaparelli's *canali* caused tremendous excitement among the public, but they remained elusive to other astronomers. Oddly, Schiaparelli initially believed his *canali* to be natural formations but was eventually won over to the canal interpretation.

Percival Lowell (1855–1916) was one of many educated enthusiasts who were captivated by the possibility of intelligent life on Mars. Lowell had been born into a wealthy, aristocratic Boston family. When he graduated from Harvard in 1876 with a degree in mathematics, his brother Abbott was president of the university, and his sister Amy was a well-known poet.

After several years of travel Lowell decided on a career in astronomy. He was intrigued by Schiaparelli's canals and was keen to explore the surface of Mars for himself. For a man of his wealth, there was no need to find a position at an observatory and start at the bottom. He would build his own observatory.

Lowell chose Flagstaff, Arizona, as the site for his observatory. Flagstaff in the 1890s was a Wild West railroad town, hardly more than a village. But at an altitude of 7,000 feet, it had a glorious view of the night sky. In 1894 Lowell commissioned the Clarks to build him a 24-inch refractor for $20,000. Two years later the telescope was brought to Flagstaff by train and installed in a dome built of ponderosa pine.

Lowell spent the next fifteen years looking at Mars and drawing its features. He was obsessed with the idea of intelligent life on Mars and wrote three books on the subject. In 1909, however, Eugène Antoniadi, a keen-eyed French astronomer, concluded that the canals were chance alignments of dark patches that the eye organizes into lines. (In recent times this explanation has lost favor because images from satellites don't show some of the dark patches seen by visual observers.)

The Clarks' next triumph was a 36-inch refractor built for the Lick Observatory on Mount Hamilton, near San Jose, California. James Lick had made a fortune in the gold rush of 1849 and invested wisely in San Francisco real estate. At the age of seventy-eight, with death approaching, Lick decided that he wanted a permanent monument to his name. He therefore transferred his $3 million fortune to a public trust, with instructions that $700,000 be donated to the University of California for the purpose of building an astronomical observatory. The university naturally chose Clark and Sons to make the achromatic lenses. A site survey settled on Mount Hamilton for the new observatory. Two years later, in 1876, James Lick died and was buried at Mount Hamilton.

The Clarks contracted with a firm in Paris to cast the huge glass blank. The blank was finally cast successfully in 1895, after five years of failures. It took the Clarks another year to grind and polish the massive lenses. First light through the 36-inch was seen in 1888.

The telescope now took its place as the largest refractor in the world. It was used extensively for photography of the Moon and the planets and made possible the discovery of many double stars. After a spectrograph was mounted to the telescope, the Sun's motion through the stars was determined to high precision, and the spectra of many nebulae were obtained.

While the Lick astronomers were waiting for their 36-inch *refractor*, they received the gift of a 36-inch *reflector* from Edward Crossley, a wealthy amateur astronomer in Halifax, England. The telescope had been designed and built in 1876 by Andrew Common, another amateur astronomer. Common's design included a number of clever innovations, such as floating the heavy polar axis on a pool of mercury to reduce friction. Later, as we shall see, this idea was applied to the great reflectors at Mount Wilson.

Common sold the telescope to Crossley, who eventually decided that the climate of England was unsuited for a telescope of this size and so donated it and its dome to the Lick Observatory in 1895. The new director of Lick, James Keeler, overhauled the telescope and put it to effective use. Charles Perinne used it to make thousands of photographs of Eros, a minor planet, as it approached the Earth to within 34 million miles in 1901. From the orbit of Eros an improved estimate of the distance to the Sun could be made. The Crossley reflector underwent various improvements over the years and has served admirably up to the present day. It has been used for pioneering work in stellar evolution, the physics of planetary nebulae, and spectral studies of faint variable stars.

The Clarks' crowning glory was the 40-inch refractor built in 1897 for the Yerkes Observatory of the University of Chicago. It remains to this day the second largest refractor ever built and represents the end of the dominance of refractors in astronomy. The telescope owes its existence to the great American astronomer George Ellery Hale (fig. 1.6).

Fig. 1.6. George Ellery Hale, a driving force in American astronomy

A FORCE OF NATURE

"He had serious psychological problems. He suffered from chronic headaches, insomnia, and frequent episodes of insanity. He had an imaginary elf, who acted as his advisor. He used to take time off to spend a few months at a sanatorium in Maine" (Richard Preston, *First Light: The Search for the Edge of the Universe,* 1987).

Despite these handicaps, which appeared only late in life, George Ellery Hale (1868–1938) had a stunning career in astronomy. He was a powerhouse of innovation and drive. No other astronomer in the first half of the twentieth century advanced astrophysics, either directly or indirectly, more than he did. His personal contributions to solar research were outstanding, but even they were surpassed by his contributions to the working tools of other astronomers.

Hale might easily have been lost to the world. He was a sickly child, surviving only though his mother's tender care. Then in 1871, at the age of three, he missed being immolated in the Great Chicago Fire only because his family had recently moved to a different part of the city.

Even in adolescence Hale was fascinated by science. His first love was microscopy, but when he was around fourteen years old his father gave him a small refractor, and from then on he was captivated by astronomy. Hale senior was a wealthy man who encouraged the boy's interest with further gifts of telescopes, gratings, and other optical equipment. One of the first instruments George built was a prism spectroscope, with which he studied the solar spectrum.

Hale attended the Massachusetts Institute of Technology and graduated in 1890 with a degree in physics. By this time he had sufficient grounding in optics and mechanics to invent his first instrument, the *spectroheliograph.* This device isolates a narrow band of the solar spectrum and generates an image of the Sun in the light within that band. With this new tool Hale could examine the solar chromosphere over the disk of the Sun and photograph solar prominences in the absence of a solar eclipse (note 1.11).

Hale was in no rush to find a job, especially since his father had offered to provide him with his own observatory at the family home. The Kenwood Physical Observatory, as it was called, opened in 1888 with a concave grating spectrograph for solar observations, a darkroom, and a shop. Three years later it was equipped with a 12-inch telescope and a suitable dome. Hale mounted his spectroheliograph to the telescope and spent every daylight hour investigating the Sun with it. He published his results in his own journal, the *Publications of the Kenwood Physical Observatory*, and in other journals of the time.

Hale might have continued working in this mode for some time, but when he learned that the University of Chicago was being organized, he got the urge to join its faculty. When he applied to the president, William Harper, he was offered a deal: he could be hired only if he agreed to transfer his observatory to the university. He declined, furious that he would not be considered on the strength of his reputation and potential alone. Within a year, however, Hale senior had closed a different deal: the university could have the Kenwood Observatory if it hired George and also committed to build a much larger observatory, valued at no less than $250,000, within two years.

Hale joined the faculty in 1892 as associate professor and, incidentally, director of the Kenwood Observatory. He immediately began to campaign for a larger telescope. Although a 12-inch instrument might be adequate for solar research, a much larger one would be needed for studies of the stars and nebulae. At a meeting of the American Association for the Advancement of Science, he happened to meet Alvan Clark junior, the optician who made the 36-inch lenses for the Lick Observatory refractor. Clark told him that he was holding the blanks for a 40-inch achromat the University of Southern California had ordered, but that the project had stalled for lack of funds.

Hale seized on this opportunity. First he persuaded the University of Chicago to buy Clark's 40-inch achromat and build a telescope with it, if they could find the funds. Then he and President Harper persuaded Charles Yerkes, a Chicago banker, to pay for a complete observatory, including an astrophysical laboratory. Yerkes's bank had collapsed in the financial panic that followed the Great Chicago Fire, and a subsequent audit showed that he had stolen $400,000 deposited by the City of Philadelphia. Yerkes was sent to prison. He spent the years after his parole attempting to restore his tarnished reputation—hence his willingness to endow an educational facility. In exchange for his donation, Yerkes had the new observatory named after him. Hale was named as its director at the tender age of twenty-four.

After five years of grinding and polishing, the Clarks finished the great achromat. The telescope and its dome were completed in 1897 by the firm of Warner and Swasey. It was a stunning engineering accomplishment for the

Fig. 1.7. The Yerkes 40-inch refractor, the pinnacle of a long line of development

time (fig. 1.7). The tube alone was 64 feet long and 52 inches in diameter, and it weighed 6 tons. The telescope was so finely balanced, however, that it moved smoothly at the push of a hand. The 90-foot dome had a 75-foot rising floor, to enable an astronomer to reach the eyepiece. The mount for the telescope was exhibited in Chicago at the World's Columbian Exposition of 1893 and created a sensation.

Hale selected Williams Bay, Wisconsin, as the site for the Yerkes Observatory. The place had great practical advantages, such as proximity to Chicago, a rail line, and dark nighttime skies. Winter temperatures in the unheated dome often plunged to 15 degrees below zero, however, which made observing a real hardship. Hale built a huge spectroheliograph with funds from the Rumford Society, hung it on the back end of the 40-inch, and used it for studies of the solar chromosphere, sunspots, and prominences. Comparing sunspot and disk spectra, he was able to detect the temperature-dependence of some spectral lines, a first step toward understanding how spectral lines are formed.

Hale assembled a staff of skilled and highly motivated astronomers at Yerkes, and they at once began to use the great telescope effectively. Shelburne Burnham used it in a thorough search for double stars. Edward Barnard, perhaps the most ardent observer on the staff, who would sit at the telescope for hours on those frigid winter nights, compiled a photographic atlas of the Milky Way and discovered the fifth satellite of Jupiter. The star that bears his name is the second-nearest to the Sun. George Ritchey continued his photographic survey of the Moon, Ferdinand Ellerman studied solar flares, and Walter Adams took spectra of faint stars. In 1903 Frank Schlesinger started a program using the great refractor to determine the parallaxes (and thus the distances) of many stars. That program continued without a break for over sixty years and yielded a model of the immediate neighborhood of the Earth.

Astronomers at Yerkes were free to pursue their individual interests, and the result was a broad, rather eclectic program. Their focus lay mostly in the solar system and among the nearby stars. There was no extragalactic astronomy as such—a surprising fact, given what was known about the universe by the dawn of the twentieth century. In 1864 Sir William Huggins, a wealthy English amateur with a private observatory at Tulse Hill, South London, had shown that spiral nebulae had a smooth spectrum overlain with faint dark lines, much like the Sun's, whereas other nebulae, like the one in Orion, emitted a spiky spectrum of bright lines that was characteristic of dilute gases. Huggins concluded that the spirals were great congregations of stars, distinctly different from gaseous nebulae. J. E. Keeler, at Lick Observatory, had obtained beautifully resolved photographs of spiral nebulae with the Crossley 36-inch. Clearly the universe was much larger than the Milky Way and was filled with "island universes," as Immanuel Kant had theorized in 1755.

Extragalactic astronomy, then, was certainly possible at Yerkes, but nothing was going on. Why? The reason might have been that even before construction of the 40-inch was completed, Hale already had plans for a larger telescope, one better suited for spectroscopy of faint nebulae and located in a milder climate. The question was whether to push for a larger refractor or to turn to reflecting telescopes. An experiment in France was crucial in helping Hale make this decision.

THE REFRACTOR REACHES ITS LIMIT

Alvan Clark and Sons were by no means the only opticians producing fine lenses. In France the firm of Paul Ferdinand Gautier had established an enviable reputation for high-quality work throughout Europe. Gautier (1842–

1909) had produced many excellent achromats, designed either for visual observations in green light or for photography in blue light. In 1885 his firm was chosen to supply thirteen telescopes for the Carte du Ciel, a massive international program to photograph the entire northern sky. The largest of these "astrographs" had an objective 33 inches in diameter and was erected at the Meudon Observatory near Paris.

In the late 1890s Gautier decided to build the largest refractors in the world, surpassing even the Clarks' 40-inch, and to display them at the Great Paris Exhibition of 1900. He cast and figured two 49-inch achromats, one for visual work, the other for photography. Each one weighed a staggering 2,000 pounds and had a focal length of nearly 200 feet. These dimensions precluded building a mount similar to that of the Yerkes telescope. Instead, Gautier built a fixed horizontal telescope for the exhibition. Starlight was directed to the objective of the telescope by a *siderostat,* a flat rotating mirror 79 inches in diameter.

Eugène Antoniadi, one of the finest astronomical observers in France, was chosen to demonstrate the power of the instrument. The stellar images he obtained were disappointing, however, partly because of an imperfect design for the siderostat drive, but mainly because of the atmospheric turbulence in the horizontal tube. Such a long tube would require careful ventilation and attention to thermal gradients.

After the exhibition, Gautier's giant refractor was moved to the Meudon Observatory, with the hope of using it for solar observations. The telescope never did perform satisfactorily, however. Gautier had invested a huge amount of money and prestige in this enterprise, and when the result was deemed a failure, his business suffered severe financial difficulties. The saga of Gautier's giant achromat convinced astronomers and opticians that refractors had reached their practical limit of size.

The Yerkes 40-inch telescope remains the largest working refractor in the world. It represents the end of a long line of development because nobody was willing or able to build larger achromatic lenses. From this point on astronomers hungry for more light from the heavens would turn to the reflecting telescope. George Ellery Hale was the pioneer in this development.

THE AGE OF HALE

GEORGE ELLERY HALE had achieved remarkable success. He had founded a new observatory that was equipped with the largest refractor ever made. He had attracted a capable staff, which was churning out valuable research papers. While waiting for the observatory to be finished, he had also founded the *Astrophysical Journal,* still one of the major journals in the field, and had become a key player in the organization of the American Astronomical Society. Such accomplishments might have crowned the ambition of most young scientists.

This energetic man was not satisfied, however. He had the vision to see that astrophysics was the wave of the future. Spectroscopy, coupled with the new physics that Gustav Kirchhoff and Robert Bunsen had revealed (note 1.9), would allow astronomers to determine the motions and composition of all the objects in the sky, a tremendous advance of knowledge. Spectroscopy, however, demands enormous amounts of additional light. In a spectrum, the light from a faint object is spread over a wide area on a film. In order to obtain a usable photograph, the astronomer needs either a larger telescope or a much longer exposure of the film.

Refractors had reached their limits in the 40-inch. Though they had their advantages, they had never been totally satisfactory, because their thick lenses absorbed the blue light to which photographic plates were most sensitive. Only great reflectors could satisfy the desperate need for more light, and Hale already had plans for such a device. In 1894 he had persuaded his father to order a 60-inch glass blank from the French firm of St.-Gobain. The disk arrived two years later, then remained in storage for years while Hale tried to find the money to build a telescope around it.

Hale first tried to interest John Rockefeller in the project. During a visit to Yerkes in 1901, Rockefeller had seen George Ritchey grinding the 60-inch blank and was impressed by the delicacy of the optical tests he was using, but the old fox would not be drawn into a donation.

Andrew Carnegie, another robber baron of the time, decided to compete with his rival Rockefeller in good works. In 1902 he announced the formation of the Carnegie Institution of Washington, which would be dedicated to encouraging research and discovery. Hale saw his chance and applied to the Executive Committee for funding for the construction of a 60-inch telescope. While waiting for an answer, he followed the progress of a survey the committee had funded, a search for a suitable site for a solar and a southern observatory. W. J. Hussey, a Lick Observatory astronomer, examined sites in California, Australia, and New Zealand. He reported excellent results at Wilson's Peak, near Pasadena, California. The 5,800-foot mountain offered clear air, rock-steady images, and a mild climate with ample sunshine. He was also impressed with Palomar Mountain, near San Diego.

Hale decided to look for himself, and in 1903 he spent three days and nights observing the Sun and stars from Mount Wilson, with Hussey and W. W. Campbell of Lick Observatory. When he left, he was convinced that the site had tremendous possibilities for a major solar and stellar observatory.

In due course the Executive Committee turned down Hale's request for funds for the 60-inch reflector but granted him $10,000 for solar work. Hale was devastated at first but resolved to continue to press for funds. Later that year he and his family moved to Pasadena so that he could make solar observations from Mount Wilson. Perhaps some test results would convince the Carnegie people. He considered this effort as just an "expedition from Yerkes," but already the idea of a permanent move to Pasadena was germinating.

Over the next year Hale built living quarters (the "monastery"), a shop, and a power shack on the mountain, using his own money when necessary. With a 12-inch coelostat (note 2.1) and a 6-inch lens he was able to take some superb photographs of the solar disk, which he used as ammunition for his cause.

After much delay, negotiation, and equivocation, the Carnegie Institute finally granted Hale $300,000 over two years. The Mount Wilson Solar Observatory was in business. The year was 1904, and Hale, at the ripe age of thirty-six, quit Yerkes to become the director.

THE SUN

Hale's first step was to haul up the mountain a horizontal solar telescope named for the donor, Miss Helen Snow. Some of the parts weighed over 300 pounds and required a special truck to carry them over the primitive trail to the summit. The great advantage of a horizontal telescope is that auxiliary instruments, like a spectrograph, can be mounted permanently and kept easily in adjustment. The telescope consisted of a coelostat, a 30-inch flat mirror, and a 24-inch concave mirror that focused an image on the slit of a spectrograph. The telescope's focal length was 60 feet and produced a solar image 6 inches in diameter.

Hale began to study the spectra of sunspots with the Snow telescope, and by comparing them with laboratory spectra he was able to show that sunspots are cooler than their surroundings. Later, Walter Adams extended these studies into a system of classification of spectral lines by temperature class, an important step in understanding stellar spectra.

In the early days at Mount Wilson, Hale had climbed trees to evaluate the *seeing*, or sharpness of images, far from the ground. He learned that warm, turbulent air near the ground distorts the solar image, and that the seeing is therefore better the higher you go. With this in mind he built a 60-foot solar tower in 1908. A coelostat at the top directed sunlight through a 12-inch lens and down to a vertical spectrograph that was mounted in a thermally insulated pit 30 feet below ground.

With this equipment Hale obtained sunspot spectra of unprecedented wavelength resolution. He discovered that some lines in the spectra were double (contrary to expectations) and, moreover, polarized. Hale guessed that these effects were caused by a magnetic field, as Pieter Zeeman had recently discovered in the laboratory. When Hale compared the sunspot spectra with his own laboratory spectra, he concluded that sunspots contain powerful magnetic fields. This was a major discovery that helped to establish the reputation of the new observatory.

THE SIXTY-INCH TELESCOPE

With Mount Wilson firmly established and funded, Hale was ready to proceed with the construction of the 60-inch telescope. In 1904 a special optics shop was built in Pasadena, and George Ritchey was put in charge of polishing the mirror. Extreme precautions were taken to maintain a constant temperature and a dust-free environment in the shop. The double windows were sealed, the air was filtered, the floor was kept wet, and the opticians wore surgical gowns while working.

The mirror blank, which had lain in storage for eleven years, was made of plate glass 7.5 inches thick and weighed 1,900 pounds. After six months, a rough concave surface had been gouged out and polishing could begin. Eighteen months later, a smooth parabolic figure was obtained, and all that remained was a final polishing. But some unknown material in the polishing compound scratched the delicate surface. Two years of painstaking work had been ruined. The mirror had to be ground down again to a spherical shape and refigured. The opticians suited up again and went back to work. Four months later, in late 1907, the mirror received its coating of silver and was pronounced finished. It was the largest glass mirror in the world.

Hale chose an equatorial fork mount for the telescope (fig. 2.1). Built in San Francisco, it had a polar axis pointing to the celestial pole and a declination axis perpendicular to it. The steel mount was massive, weighing 22 tons in total. The polar axis alone weighed 4.5 tons and was 15 feet long, while the cast-iron fork that held the open telescope tube weighed 5 tons. Despite its weight and size, this huge mass had to rotate perfectly smoothly to follow a faint object for hours.

Hale recognized that the great weight of the telescope would place an extreme load on ordinary bearings. He had to find a novel design solution. Astronomers had previously experimented with the idea of floating a telescope's polar axis in a pool of mercury. The mercury would not only support the telescope but also provide a practically frictionless bearing. The idea was attractive, but nobody had made it work. Nevertheless, Hale and Ritchey decided to try once again. They built a steel float 10 feet in diameter and attached it to the polar axis. The axis then floated snugly in a steel tank filled with 650 pounds of mercury. With a narrow clearance between the float and the tank, most of

Fig. 2.1. The 60-inch telescope at Mount Wilson Observatory

the telescope's weight was supported, and mechanical bearings had to take up only a small fraction of the load. Hale had overcome another obstacle, but there was more to come: In April 1906, when the mount was nearly complete, San Francisco suffered its Great Earthquake and the huge fire that followed. Yet amid all the devastation, Hale discovered that the mount and the mirror for the 60-inch had survived. He set to work rebuilding his shops and within a few months was back in business.

A few major parts still had to be finished. The huge gear that drives the telescope, 10 feet in diameter and weighing 2 tons, had its teeth cut in Pasadena and was then fitted to the mount. The telescope, without its mirror, was then assembled for a mechanical test, which it passed with flying colors.

Hale now faced the problem of getting this ponderous machine to the top of Mount Wilson. The mount was taken apart and crated, but before the telescope could make the trip the narrow, winding trail up the mountain had to be widened and straightened. Hale took on the role of civil engineer, and in 1907, after a year of roadwork, the shipment was ready.

A special truck was built to haul the load. The truck had a high-torque electric motor on each wheel, and the front and back axles could be turned independently, to get around the sharp turns in the road. The truck was used to carry the heaviest pieces to the observatory, but mules hauled most of the parts, including the segments of the dome. A total of 150 tons of equipment was carried to the top by teams of mules, a tremendous task indeed.

At the top, the telescope was reassembled on its pier, and a 58-foot dome was built around it. On December 7, 1908, the great mirror was installed, and the telescope was complete at last.

SHAPLEY AND THE GALAXY

The 60-inch was put to work immediately to obtain the spectra of distant nebulae. The Andromeda Nebula, Messier 31, had a spectrum similar to that of the Sun, which implied that it consisted of Sun-like stars. In fact, individual stars and clusters could now be resolved in the Andromeda as in other spirals. The study of our own Milky Way really began, however, with the work of Harlow Shapley.

Shapley (1885–1972) was a student of the great Henry Norris Russell at Princeton. He completed his doctoral dissertation on eclipsing binary stars in 1913 and joined the staff at Mount Wilson the following year. Shapley had just learned about a remarkable discovery made by Henrietta Leavitt at the Harvard College Observatory. The observatory's director, Edward C. Pickering, was paying Miss Leavitt 30 cents an hour to search for *variable stars* (stars whose brightness regularly increases and decreases) on the photographic plates that had been obtained at the Harvard Station in Peru. She concentrated her attention on the Magellanic Clouds, which are satellite galaxies of our Milky Way Galaxy. In these Clouds she discovered over seventeen hundred variable stars. Among these, she found twenty-five *Cepheid variables,* yellow supergiant stars that varied in brightness in periods between one and six days. In 1907 she announced that, among these Cepheids, the longer the period, the brighter the star. She realized that because these stars were in the Clouds, at approximately the same distance from Earth, the absolute brightness of a Cepheid is related to its period. She had discovered the *period-luminosity law* for Cepheids.

Shapley realized that this relationship offered an excellent tool for determining the distances of nebulae. But first someone had to calibrate the law by finding the distance of at least one Cepheid. Ejnar Hertzsprung, a Danish astronomer, found thirteen Cepheids in our galaxy and used the standard technique of statistical parallax (note 2.2) to determine their distances and therefore their intrinsic brightness. Shapley improved on Hertzsprung's method, then in 1914 began using the 60-inch telescope to search for Cepheids in the so-called globular clusters within the Milky Way. These are tight clusters of up to one hundred thousand stars, among which are some of the oldest stars. Shapley found Cepheid-like variables with periods shorter than a day, assumed that they followed the same law as the classical Cepheids in the Magellanic Clouds, and derived their distances from their periods.

The result was astounding. Shapley discovered that the globular clusters form a huge spherical halo around our disk-shaped Milky Way. He estimated that our galaxy is 300,000 light-years in diameter, about ten times larger than previously estimated, and that our solar system lies far on the outskirts of the galaxy. Shapley's assumptions, and hence his derivations, were in fact mistaken (note 2.3). We now know that the Milky Way has a diameter of about 100,000 light-years and that our Sun lies about 26,000 light-years from the center.

Shapley's conclusion was too radical to be accepted immediately. It amounted to a second Copernican Revolution, in which the Sun and Earth were once again demoted from a special status in the universe. In 1920 Shapley debated Heber Curtis, a Lick astronomer, on the scale of the universe at a meeting at the National Academy of Sciences. Curtis argued that the Milky Way is only one of many spiral galaxies in the universe, and that the Sun lies at its center. Shapley, in contrast, maintained that the Milky Way comprises the entire universe, that the many spirals astronomers had discovered are gas clouds that lie within our supergalaxy, and that the Earth lies far from its center. In hindsight we know that both astronomers were partially right and partially wrong. But a resolution to the issue would have to await the construction of a larger telescope.

THE ONE-HUNDRED-INCH TELESCOPE

Hale was thinking about a larger telescope well before the 60-inch was completed. He had been introduced to John Hooker, a rich Pasadena manufacturer of hardware, and tried repeatedly to interest the man in funding a large telescope. Like Yerkes and Carnegie, Hooker yearned for some philanthropic project that would improve his somewhat dubious reputation—but only at a reasonable price. Hale spent many evenings with the Hookers talking about astronomy, and many afternoons with the lovely Katherine Hooker in her Italian garden.

Eventually Hale's persistence paid off. Hooker asked the price of an 84-inch telescope mirror, and Hale told him $25,000. Hooker thought it over and surprised Hale by offering him $45,000 for a 100-inch, with the condition that his name be attached to the finished telescope. As usual, Hale would have to find additional funds to complete the telescope. Hooker was nothing if not a shrewd dealer.

In September 1906, two years before the 60-inch would be finished, Hale ordered a 100-inch blank from the same French firm, Plate Glass Works of St.-Gobain, that had cast the 60-inch blanks. The mirror would require 4.5 tons of glass to be cast, more than the firm's furnaces could melt at one time. So the blank was poured in three sessions and allowed to anneal for a year. When it finally arrived, Hale and Ritchey were horrified. The three layers of glass hadn't fused properly, and Ritchey concluded that the blank could never be ground to a final mirror.

Hale persuaded St.-Gobain to cast a second blank, at its own expense. The firm constructed a larger furnace and annealing oven especially for this mirror. In 1910 a second blank was cast, also in three layers. This time air bubbles were trapped between the layers, which weakened the great glass disk. The blank cracked during the annealing process and had to be scrapped.

Hale was at his wit's end. Hooker's original $45,000 had been spent, and a usable blank still seemed years away. But at this point Walter Adams, a senior astronomer at Mount Wilson, sent good news. Contrary to Ritchey's opinion, the first blank was indeed strong enough to bear grinding into a parabolic shape. And, to crown the occasion, Andrew Carnegie had decided to give his Institution an additional $10 million, with special instructions for funding the completion of the 100-inch telescope.

Ritchey grudgingly undertook the grinding of the original 100-inch blank. If he had had his way, he would have scrapped it, but Hale was adamant. Despite its flaws, the great disk, 13 inches thick and made of the green glass commonly used for wine bottles, held together.

The telescope mount was designed by a mechanical engineer, Frances S. Pease, with advice from Hale and Ritchey. A firm in Massachusetts built the great steel structure while the mirror was being figured in Pasadena. The equatorial mount (fig. 2.2) employs a huge rectangular yoke that rotates on an axis

Fig. 2.2. The 100-inch Hooker Telescope at Mount Wilson Observatory

parallel to the Earth's axis. The telescope tube and mirror pivot in declination about an axis perpendicular to the yoke. Together, the tube and mirror weigh 87 tons. To reduce the friction on the support bearings, a mercury float system was installed, similar to the one in the 60-inch. A special mechanical clock, built in the Pasadena shops, governs the tracking of the telescope, which must be accurate to one-tenth of an arcsecond.

The fragile mirror posed a special problem. Despite its thickness, a mirror this size will flex under gravity as the telescope turns. To offset this problem, a special mirror cell was designed to support the huge glass disk. This cell was critical, and the problem cost Hale many sleepless nights until a prototype could be tested.

Once the mount and telescope tube were ready, they were shipped in pieces from Massachusetts to San Pedro by way of the Panama Canal. Hale again faced the difficult task of transporting a massive mount, a precious mirror, and the parts of a 500-ton dome to the top of Mount Wilson, where the telescope pier was waiting. Once again the road was widened and a special truck was ordered. Once again the main task fell to the mules and their drivers.

This time, however, the dome was built first and the telescope was assembled inside. The final touch was the cutting of the 18-foot worm gear that turns the telescope as it tracks a star across the sky. Any error in the spacing of the gear teeth would translate into a periodic tracking error, which would blur a stellar image. A famous Italian instrument maker, C. Jacomini, was brought in to cut the gear teeth with the gear installed on the telescope axis. Using a microscope, he divided the circumference of the gear into 1,440 equal parts and cut each tooth separately.

Hale and Adams had their first look through the completed telescope on November 1, 1917. Once the mirror cooled down to the outside air temperature, it offered a spectacular view of the star Vega. Hale was overjoyed. The project had taken over eleven years, but the result was worth the wait.

SUNSPOTS

Hale had not been idle while waiting for the 100-inch to be finished. He had used the 60-foot solar tower to learn that sunspots contain powerful magnetic fields. Moreover, he discovered that the leading spots in the northern and southern hemispheres of the Sun have opposite polarities (note 2.4). Now he conceived the idea of trying to relate changes in the Sun's magnetic fields to the magnetic storms that occasionally swept the Earth. For this purpose he built a 150-foot solar tower, which would offer a larger and more stable solar image. Beginning in 1912, Hale and his assistants measured the polarity and strength of sunspot magnetic fields, following their changes day by day.

Hale was lucky. In 1912 a new solar cycle was beginning, and a few pairs of spots of the new cycle appeared at high solar latitudes. Hale noticed immediately that the polarities of the new pairs were opposite to those of the old cycle. In the northern hemisphere the leading spots were now of northern polarity, the reverse of the situation in the period from 1908 to 1912. This effect persisted and strengthened as the months went by. Hale had discovered another fundamental property of the solar cycle.

SLIPHER

In 1917 the Mount Wilson Observatory possessed the two largest telescopes in the world, the 60-inch and the 100-inch. In that same year, an astronomer at the Lowell Observatory was making history with a puny 24-inch telescope, the Clark refractor.

As we saw in the previous chapter, Percival Lowell was obsessed with the possibility of intelligent life on Mars. He spent much of his time making detailed drawings of the surface of the planet and promoting his idea that the ruins of a great civilization were visible in the form of its canals. When Vesto Slipher (1875–1969) joined the Lowell Observatory in 1901, Lowell put him to work photographing Mars, hoping that objective photos would support his assertions. As luck would have it, the photographs helped to discredit the idea.

Slipher began taking spectra of the planets and from these determining their rotation periods. Then Lowell had another idea. Spiral nebulae, he thought, were actually gas clouds in the process of forming solar systems like our own. He wanted Slipher to measure their rotation periods so as to bolster this hypothesis.

Slipher set to work with the 24-inch telescope and a new spectrograph. His first target was the Andromeda Nebula, Messier 31, because it is large and bright. This was an unfortunate choice, in a way. The plane of M31 is strongly inclined to the line of sight, so that the arms are crowded together. Nevertheless, Slipher obtained good spectra of the whole nebula. He compared the wavelengths of many spectral lines with their counterparts in laboratory sources, and by measuring the wavelength shifts (see note 2.5 on the Doppler effect) he determined that M31 is approaching Earth at a speed of 300 km/s. Intrigued by this result, he extended his program to include fainter nebulae. By 1915 he had assembled the spectra of fifteen nebulae. Some of these were so faint as to require exposures of eighty hours with Lowell's 24-inch telescope.

Then Slipher noticed a curious phenomenon. He had expected to find that the nebulae were moving through space in random directions. If so, roughly half of his sample should be approaching Earth and half receding from it. But he found instead that eleven of the fifteen were receding. Their velocities, some as high as 1,100 km/s, were very far from being random, a most peculiar result.

He published his work in 1915, and the world ignored it—everyone, that is, except for Edwin Hubble.

HUBBLE

Edwin Hubble (1889–1953) had some difficulty choosing a career. He was a fine amateur boxer in college and might have made a living at it. After graduating from the University of Chicago with a degree in physics, he studied Roman and English law at Oxford. He returned to America in 1913, passed the bar exam in Kentucky, and practiced law half-heartedly for a year in Louisville. But the law didn't satisfy him, and he decided to chuck it all for astronomy. He enrolled at the University of Chicago and obtained his doctorate but then enlisted in the infantry to serve in France during World War I. He returned in 1919 as a major and took up Hale's invitation to join the staff at Mount Wilson.

Hubble had acquired a mustache, a pipe, a tweed jacket, and English manners. For all his affectations he was a hard worker. He began looking for novas (exploding stars) in spiral galaxies, first with the 60-inch telescope, later with the 100-inch. At the time, the status of the spirals was highly uncertain. Some, like Harlow Shapley, thought they were collections of stars within our Milky Way Galaxy. Others, like Heber Curtis, thought they were distant "island universes." Hubble thought that if he could find novas in some of the spirals and compare their brightness to novas in the Milky Way, he might determine the distances to the spirals.

He began with the Andromeda Nebula, M31, because he suspected it might be the nearest to our own. The great light-gathering power of the 100-inch tele-

scope enabled him to resolve faint stars at its edges. He took hundreds of photographic plates of Andromeda, looking for sudden changes in a star's brightness that would mark it as a nova. By 1923 he had found a few novas and many
variable stars. Among these he discovered a Cepheid variable from its distinctive light curve, a steep rise and gradual fall.

Here was a tool with which Hubble could determine the distance of the
nebula. He used Shapley's period-luminosity relationship and, to his delight,
discovered that M31 lay at an enormous distance from Earth, about a million
light-years. It had to be a distant galaxy as large as our own.

This was a discovery of the first rank, but before he could announce it, Hubble needed confirmation of his result. So he examined other nebulae, including Messier 33 in Triangulum, searching for more Cepheids. He found twenty-
two in M33 and twelve in M31. They fit a period-luminosity relationship
nicely. The first nebula "definitely assigned to a region outside the galactic system," however, was NGC 6822, at a distance of 700,000 light-years. In the
decade from 1926 to 1936, Hubble measured the distances to many galaxies and
showed that they were distributed uniformly in all directions and distances.

Hubble's work opened a new era in astronomy. He had proved that the universe is much larger than Shapley, for one, had ever imagined and that it is
filled with millions of galaxies like our own. Now many questions arose. What
is their distribution in space? Why do they differ in shape and mass? What lies
between them?

Hubble embarked on an ambitious survey of hundreds of galaxies. In time
he noticed gradations in their structure. Some were tight flat spirals, like the
Triangulum Nebula, others had no spirals but only a smooth elliptical shape,
and still others had a loose, open appearance with no definite organization. He
constructed a system for classifying galaxy shapes, which he suggested represented an evolutionary sequence.

Hubble was aware of Slipher's puzzling results on wavelength shifts, and in
the light of his own discoveries he began to see the need for a systematic at-
tack on the question of the distances of the nebulae. He would search for
Cepheids, for novas and for any other criteria, such as size and brightness, to
estimate the distances of nebulae. It was a huge task, for which he needed help.
He enlisted the aid of Milton Humason, who had worked his way up from
mule driver to assistant observer at the observatory. They shared the labor of
observing for almost ten years before they had a definite result. Fortunately

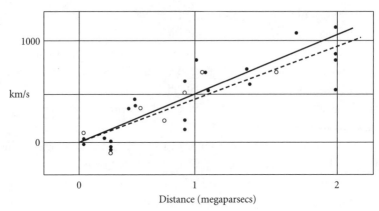

Fig. 2.3. Edwin Hubble's diagram of the relation between the distance and speed of extragalactic nebulae

they had almost unlimited time on the 100-inch telescope to carry out their program. In 1929 Hubble published his first diagram relating the measured distances and redshifts or velocities (fig. 2.3). The data were sparse but very convincing. The greater the distance, the greater the speed. The slope of the line in the diagram, now called the Hubble constant, implied a velocity gradient of about 500 km/s per megaparsec (a *parsec* equals 3.26 light-years, and a *megaparsec* equals a million parsecs). In an adjoining article, Humason announced that the nebula NGC 7619 has a recession speed of 3,779 km/s, far greater than any that had been seen previously.

We now interpret Hubble's redshift-distance relation as evidence that the universe is expanding. Indeed, as early as 1922 Alexander Friedmann, a Russian cosmologist, had used Einstein's general theory of relativity to construct model universes that *had* to expand. But Hubble was no theoretical cosmologist, and neither was Hale. If anything, Hubble preferred the static universe predicted by the Dutch cosmologist Willem de Sitter. He offered a rather vague explanation of his redshifts in terms of de Sitter's strange spectral predictions.

In any case, our ideas about the universe were changed forever. By 1936 Hubble and Humason had measured a recession speed of 40,000 km/s in a galaxy so faint as to be almost at the limit of the 100-inch. Hubble needed more light to extend his spectroscopy deeper into space, and that meant a larger telescope.

THE PALOMAR TWO-HUNDRED-INCH

Hale was overjoyed by Hubble's progress and saw clearly the need for the next giant telescope, something in the range of 200 to 300 inches. He estimated the cost of a 200-inch mirror at $500,000 and the cost of a complete telescope at $6 million. Where could he find such a sum? This was in 1928, long before the federal government had become a possible source of funds. Hale approached officials of the Rockefeller Foundation informally. After more than a year of subtle campaigning, he finally persuaded them to fund a 200-inch telescope. It would be owned and operated by the California Institute of Technology.

The next questions were technical. Who could cast such a mirror? What would it be made of? The French firm of St.-Gobain had encountered serious problems in casting a 100-inch glass blank. A 200-inch would be no easier.

Hale always kept up with the latest in technology. He now remembered hearing about fused quartz, a new material that engineers at the General Electric Corporation were experimenting with. Quartz is intrinsically stronger than glass, but more important, it hardly expands when heated. In fact, it's possible to play a blowtorch on a disk of fused quartz without cracking it. A large mirror blank made of fused quartz would anneal faster and could be ground and polished more aggressively than a glass one. The finished mirror in the telescope would also remain unaffected by temperature changes from day to night.

Hale contracted with GE to cast a series of fused-quartz mirrors, up to a size of 200 inches. Over the next three years, and at cost of over $600,000, the GE engineers struggled to fill his order but without success. Instead of casting a blank, they tried to spray molten quartz onto a base. The result was often a stack of improperly fused layers. They finally produced a 60-inch blank, but it had flaws and bubbles.

In desperation, Hale turned to the Corning Glass Works in New York State. Corning had perfected Pyrex, a type of glass designed for home cookware. Pyrex has a low coefficient of expansion—not as low as fused quartz, but adequate. A Pyrex dish can be removed from a freezer and placed directly in a hot oven without cracking. It had exactly the properties Hale needed for his big mirror.

Corning had never produced an astronomical mirror of Pyrex, although several observatories had expressed interest in purchasing one. The astronomical market was too small to encourage Corning to invest capital for new equipment, and so its original prices were outrageous. With the onset of the Great Depression, however, Corning was hungry for new business.

In 1931 Hale struck a deal with Corning. The company would produce a 30-inch, a 120-inch, and finally a 200-inch mirror. No definite price was set. Instead, Corning would get a guaranteed 10-percent profit over the cost of manufacture. The engineers agreed to adopt an idea that had arisen during the fused-quartz fiasco. Instead of a solid disk several feet thick, they would cast a thinner blank that had supporting ribs on its backside. The blank would weigh far less, and would adjust to changes in temperature more rapidly, than a solid disk.

The ribs were formed by assembling ceramic cores in the mirror's mold. Glass was poured into the mold and trickled down between the cores to form the ribs. But the cores were buoyant and tended to bob up in the glass. Corning engineers experimented with a variety of restraints until they learned that steel bolts would work. They also learned that the glass had to be heated to an usually high temperature, 2,700°F to allow it to flow between the ribs. That required building a special furnace.

In May 1932 Corning cast a perfect 30-inch blank. The following year a 120-inch blank was cast, and after six months of annealing the opticians found that it had only a few minor flaws. A 200-inch blank now seemed feasible.

Corning proceeded to build a larger furnace and an annealing oven. In May 1934, 65 tons of glass were melted and held ready for pouring into the mold. Glass was carefully ladled into the mold, 750 pounds at a time. The pouring process seemed to go well, until the workmen discovered that some of the cores had come loose and were floating in the molten glass. The disk was allowed to cool rapidly to the point where workmen could remove the cores by sandblasting. The disk was remelted to reform the scarred surface and then annealed. When at last the disk had cooled to room temperature, the engineers discovered it was still too flawed to be useful. Corning agreed to replace it at no cost, to preserve its commercial reputation.

The second 200-inch disk was also plagued with troubles. The pouring went well this time, but an unprecedented flood threatened to interrupt the power supply to the heaters during the annealing process. Then a mild earthquake shook the town of Corning. The disk was saved, but when it emerged from the annealing oven it too was found to have deep scars on its surface, apparently caused by contact with the roof of the oven. Ultimately, Hale and his advisers agreed to grind off the disk's top 2 inches to remove the flaws. The 20-ton disk was crated in March 1936 and shipped to Pasadena by rail for grinding and polishing. The trip lasted two weeks because the train traveled at only 25 miles an hour and stopped each night on a siding, protected by armed guards.

Meanwhile the design for the telescope mounting was completed. The main mirror and its secondary would be held in a massive open truss, which pivoted on a declination axis between the arms of a gigantic fork (fig. 2.4). The fork was supported on piers at both ends of the polar axis. A huge horseshoe bearing at the north end of the fork bore the 500 tons of the mount.

The telescope would have observing stations at three different foci. Light

Fig. 2.4. The 200-inch Palomar giant

from the main mirror would come to prime focus at the top of the truss. This telescope was so big that the prime focus was accessible to an observer who could crouch in a tiny cage. With a 60-inch secondary mirror in place, the telescope beam would be reflected down the truss through a hole in the primary mirror and to the Cassegrain focus. A spectrograph or photometer could be attached there and would ride along with the telescope. Finally, by inserting a third mirror, the beam could be sent to a fixed Coudé focus, where a large heavy spectrograph could be mounted permanently. (A *Coudé focus* is an auxiliary focal point, produced with a convex secondary mirror and one or more flat diagonal mirrors. The design is intended to permit heavy instruments to be mounted to the telescope without disturbing its motion.) Hale had a difficult time finding a contractor capable of building the huge mount. He eventually contracted with Westinghouse, a firm that had experience building enormous ship turbines.

A special optics shop was built in Pasadena to grind and polish the great mirror. After a year devoted to grinding off the top 2 inches of glass, the opticians hoped to find all strains and fractures gone. But the remaining blank was still heavily flawed. Corning began to discuss the possibility of casting a third blank, using a new process. Ultimately the opticians decided to proceed with the second blank, and the idea of a third blank was shelved in the summer of 1937. By 1938 a spherical surface had been ground into the blank and figuring was to begin. Parts of the great mount were hauled up to the summit of Mount Palomar in a caravan of trucks. The huge dome was nearly complete.

Hale's health had been deteriorating for several years, and on February 21, 1938, the great man died. The giant telescope project would have to be completed by Walter Adams, the new director. A target date of January 1, 1942, was set for completion.

But the events of December 7, 1941, changed all that. World War II effectively stopped the telescope project. Critical men went off to work in the war industries, and Westinghouse shifted to wartime production. The wiring for the mount controls and the polishing of the great gear would have to wait.

Upon the conclusion of the war in 1945, work on the telescope was resumed. In June 1948 the telescope was dedicated at a public function on the mountain. The project had taken twenty years, but the largest telescope in the world was now ready for service.

Humason picked up the program of the redshifts of the galaxies where he

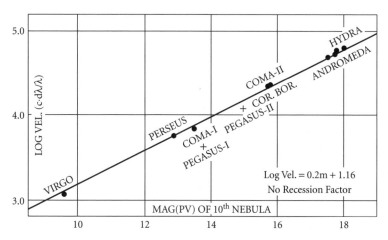

Fig. 2.5. Edwin Hubble's diagram summarizing the relationship between the logarithm of the recession velocity and the apparent magnitude (a logarithmic measure of brightness) of distant galaxies. Fainter nebulas recede faster.

and Hubble had left it before the war. By 1956, using the 200-inch, Humason and co-workers had detected a galaxy receding with a speed of 60,000 km/s. They might have probed even deeper into space were it not for a limiting factor: the night sky glows with a faint light emitted by the Earth's upper atmosphere. The fainter galaxies Humason wanted to pursue were lost in this almost invisible glow. In 1953 Hubble summarized their work in his Darwin lecture, the crowning moment of his career. Figure 2.5 shows their results.

The giant telescope on Palomar Mountain has continued to produce cutting-edge science to the present day, as we shall see in the following chapters. However, new techniques and bigger telescopes have allowed astronomers to push ever deeper into space to extend and refine Hubble's famous relationship. The Hubble constant is still one of the most important in cosmology.

WIDE-EYED ON PALOMAR

In the midst of his struggles to complete the 200-inch telescope, Hale learned about a different breed of telescope and decided that he wanted one.

A large telescope, like the 200-inch, excels at viewing faint, distant objects. Its field of view, however, is usually limited to about 30 arcminutes, or about the size of the full Moon. For most purposes this is no disadvantage, because the astronomer is usually interested in an object (a single star, a star cluster, or a galaxy) much smaller than the Moon. But to map large areas of the sky, to find particular objects, for example, it helps to have the largest possible field of view. Bernhard Schmidt (1879–1935), a talented Estonian-born optician, showed how to manage that.

Schmidt lost his right hand and forearm in an accident at age fifteen. Despite this handicap he trained successfully as an optician and established his own small workshop in Germany in 1904. His reputation as a careful craftsman spread gradually. He was able to earn a modest living with commissions for mirrors from amateur astronomers and even a few professionals. But in 1927 his business failed, and he was forced to take a job at the observatory at Hamburg. There he learned what astronomers wanted in the way of optics. Around 1929 he conceived the idea of his famous telescope.

A parabolic mirror is ideal for focusing a star to a pinpoint, but it doesn't work nearly as well on an extended object. Rays of light that enter the paraboloid at an angle don't come to the same focal point as those that enter parallel to the axis. Schmidt realized that he could correct this kind of aberration (note 2.6) if he inserted a specially ground correcting plate at the entrance of the telescope. In fact he didn't need a paraboloid at all but could use a spherical mirror and correct all its aberrations with a single plate. The plate is thickest at the center and edges and thinnest in between (fig. 2.6).

Schmidt built his first telescope in 1930 with a 17-inch mirror and a 14-inch plate. It had the very fast focal ratio of f/1.75 and a field of view of 7.5 degrees,

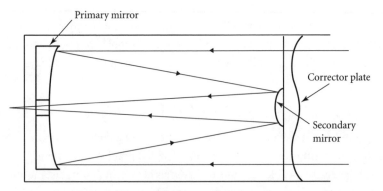

Fig. 2.6. In the Schmidt-Cassegrain, a thin aspheric plate corrects the aberrations of the spherical primary mirror. The secondary mirror reflects the light to a focus behind the primary.

or fifteen times the size of the Moon. (The *focal ratio* is the mirror's focal length divided by its diameter.) In this first version the focus was inside the telescope, behind the plate, on a curved surface. In a more useful version, the Schmidt-Cassegrain, a small secondary mirror behind the plate passes the light through a hole in the primary mirror and forms a final image behind it.

George Ellery Hale learned of Schmidt's invention from Walter Baade, one of the most famous Mount Wilson astronomers, who had discussed the design with Schmidt back in 1929. Hale commissioned a copy of the 17-inch version, which was put into service at Palomar around 1936. This instrument worked so well that a 48-inch version was started in 1939 and finished only in 1948. With a focal ratio of f/2.5 and using 14-inch photographic plates, it has a field of view of 7 degrees. It was used to carry out a complete mapping of the northern sky, in red and in blue light. By comparing images of the same star in the two colors, an astronomer can determine its surface temperature. These two-color surveys have served as an invaluable reference tool ever since.

During the three decades between 1950 and 1980, the science of astronomy advanced on at least three fronts simultaneously. In the immediate postwar period, astronomers were able to view the universe for the first time at x-ray, infrared, and radio wavelengths, with tremendous consequences. Second, major astronomical centers were established at several prime sites and equipped with optical telescopes as large as 160 inches, or 4 meters, to probe deeper into space than ever before. And in parallel with these developments, two revolutions occurred in the art of making mirrors. These technical advances set the stage for the construction of the giant telescopes of the 1990s and the new millennium.

NEW WINDOWS ON THE UNIVERSE

DURING THE three decades that followed the end of World War II, astronomy was utterly transformed. For three hundred years astronomers had been able to view the sky only in visible light. They knew that the light we can see constitutes only a part of the whole available spectrum, but they had no idea that stars might emit x-rays or radio waves. Even if they had entertained such a thought, they didn't have the technology to test it. As figure 3.1 shows, light from the universe leaks through the Earth's atmosphere in only a few narrow windows. Astronomers had been able to look through only one of these.

World War II changed all that. Technical advances made during the war allowed astronomers to open all the windows and led to a veritable revolution in our ideas about the universe, our world-view. Rockets left over from the war enabled researchers to lift their instruments above the atmosphere, if only briefly, to sample x-rays from space. And the explosive growth of electronics during the war led directly to the new discipline of radio astronomy. Lagging somewhat behind were advances in our ability to detect and make images in infrared radiation. One might also mention the advent of computers, now an essential tool of the astronomer, as a technical spin-off from the war.

In three decades between 1950 and 1980, a flood of discoveries revealed a universe more violent and more bizarre than any imagined earlier. A new field of study, high-energy astronomy, raised challenging questions about the origin of the universe, the evolution of stars and galaxies, and the properties of interstellar space. Optical astronomers—the good old-fashioned kind—were presented with a broader context and a new agenda.

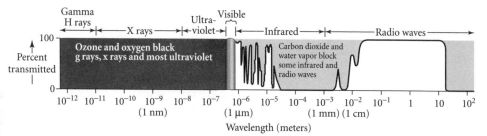

Fig. 3.1. Water vapor, ozone, and such gases as carbon dioxide in the Earth's atmosphere block out much of the light that arrives from space. This graph shows the places in the spectrum where atmospheric transmission is high.

BROADCASTS FROM THE SKY

It all began with a happy accident. Karl Jansky, a physicist working for Bell Telephone Labs, was searching for the source of the static, with wavelengths from 10 to 20 meters, that was interfering with radio-telephones. Jansky cobbled together a radio antenna with which to search the sky. He discovered that thunderstorms, both near and far, produced most of the static. But in addition there was a steady hissing noise that seemed to move across the sky.

At first Jansky thought the source was the Sun, but when a partial eclipse in August 1932 failed to change the signal, he decided to look further. By December he had concluded that the source was traveling across the sky at the same rate as the stars. Finally, in 1933, he pinned it down. The center of the Milky Way, in the constellation of Sagittarius, was the culprit. Jansky had discovered that our galaxy emits radio waves as well as light.

Although Jansky's discovery made the May 5, 1933, edition of the *New York Times,* neither Bell Labs nor the rest of the scientific world took any notice. Grote Reber, a radio engineer in Chicago, was the sole exception. After trying unsuccessfully to work with Jansky at Bell Labs, he decided to build a radio telescope in his backyard using his own money. It was the prototype for all succeeding telescopes, a parabolic dish 10 meters in diameter, fully steerable in an alt-azimuth mount, with a narrow-band radio receiver at its focus.

After several failures, Reber detected emission from the Sun, from the center of the Milky Way, and from the direction of the constellation Cassiopeia. He settled down to a systematic survey of the whole sky at a wavelength of 1.9 meters. In 1944 he published the first maps of radio emission in the Milky Way, which showed strong sources in Sagittarius, Cygnus, and Cassiopeia. Reber pointed the way toward a new astronomy, and after World War II other scientists were free to follow his lead.

MAPPING THE RADIO SKY

Reber's discoveries raised a host of questions. What was the source of the galaxy's radio emission? Was it a smooth sea of gas between the stars? Or did all the stars emit radio waves? If so, how did they produce their radio emission?

John Bolton was among the first to tackle these questions. He was a Yorkshireman who had spent the war working on radar. After the war he joined the Radiophysics Laboratory in Sydney, Australia, and was given a chance to use some homemade antennas originally designed to observe a total eclipse of the Sun. Bolton moved the gear to the top of a high cliff overlooking the sea. Earlier, Joseph Pawsey of the Radiophysics Lab had shown how to use the sea as one arm of an interferometer (note 3.1) in order to achieve higher angular resolution with a small antenna. With this setup, Pawsey discovered that sunspots radiate at meter wavelengths. Bolton and his colleagues Gordon Stanley and O. Bruce Slee adopted Pawsey's method to search for radio stars.

In 1948 they rediscovered the strong source in Cygnus, which they labeled Cygnus A. The source was very narrow, less than 8 arcminutes across, so they guessed it might be a star. Later they found two more strong sources, Virgo A and Centaurus A, but were unable to identify them with any known objects. They had better luck with Taurus A, which turned out to coincide with the famous Crab Nebula, a supernova remnant. But to make further progress they had to nail down the locations of the sources more accurately. Their sea interferometer was not up to the task.

In England, Martin Ryle was getting into the field. Like Bolton, he had worked during the war as a physicist to develop radar. After the war he settled at Cambridge University and began to gather a small group of like-minded physicists interested in using their skills in a new field. He also recognized the importance of getting accurate positions for the puzzling new radio sources and, like Bolton, decided that an interferometer was the optimum tool. So he built one that consisted of two antennas on an east-west baseline and a number of receivers sensitive to meter wavelengths. The Earth's rotation allowed him to scan the entire northern sky with a narrow fan-shaped beam. (We'll discuss interferometers in much more detail in chapter 10.)

In 1951 Ryle's colleague Francis Graham Smith obtained a position for Cygnus A accurate to 15 arcseconds and alerted Walter Baade and Rudolph Minkowski at the Mount Wilson Observatory (note 3.2). These two famous astronomers rushed to the 200-inch (5-meter) telescope, where they discovered something extraordinary. Cygnus A was not a star, but a faint galaxy, apparently with two nuclei. They concluded that the source was in fact two galaxies in collision. (They would turn out to be wrong.) From their redshifts Baade and Minkowski determined that their distance was a staggering 1 billion light-years. How, they wondered, could such distant galaxies produce a strong radio signal while nearby stars were undetectable?

Bernard Lovell, another former radar physicist, was also beginning to search for radio sources at meter wavelengths. He built a cheap antenna at the isolated English village of Jodrell Bank. This 66-meter "wire bowl" was immovable, but by shifting his receiver Lovell could reach different altitudes above the horizon, and the Earth's rotation would then allow him to scan east-west strips of the sky.

In 1951 Robert Hanbury Brown and Cyril Hazard, two of Lovell's associates, used this telescope to show that the Andromeda Nebula, a galaxy like our own, was a radio source at 1.89 meters. By 1953 Hanbury Brown and Hazard had found twenty-three discrete radio sources, but without any clear identifications with other nebulae.

Jodrell Bank was now competing with other groups to find radio sources. John Bolton and Bernard Mills in Australia and Martin Ryle and friends at Cambridge University had, among them, cataloged over a hundred of these sources. It was becoming clear that the sky might be filled with discrete radio sources. Lovell realized that he would need a more powerful telescope. So he campaigned successfully to find the funds to build a 76-meter steerable dish, which remains the second largest in the world. He and his colleagues began a systematic search for sources at meter wavelengths.

At Cambridge University, Ryle established another major center for radio astronomy. Starting in 1950 and continuing for over twenty years, Ryle and his associates (principally Antony Hewish) compiled a series of catalogs of sources, the 1C, 2C, and up to 7C. The 3C catalog, for example, contains almost five hundred sources, which include many galaxies and supernova remnants.

With more sensitive receivers, Ryle and Hewish detected over five thousand sources.

THE MILKY WAY'S ARMS

When Jan Oort, the great Dutch astronomer, learned of Grote Reber's discovery of radio emission, he recognized its potential for probing the Milky Way. Unlike visible light, radio waves would not be scattered by interstellar dust clouds (note 3.3), and they might allow a clear view into the central bulge of the galaxy. Radio waves would be especially valuable if they were concentrated into narrow bands of frequency, in analogy to the atomic spectral lines in visible light. If such lines existed, Oort thought, their Doppler shifts might be used to detect the motion of interstellar gas clouds.

Oort gave his graduate student, Hendrik van de Hulst, the task of searching for such narrow bands in the radio spectrum of hydrogen, the most abundant element. In 1944 van de Hulst predicted a line at a frequency of 1,420 MHz or a wavelength of 21 cm (note 3.4). But with the war on and with the chaos that followed, the Dutch were unable to follow up on his prediction.

Not until 1951 was a test possible. In that year Edward Purcell, a professor at Harvard University, and Harold Ewen, his graduate student, built a special receiver and a small horn antenna, set them up outside a window of the Lyman Physics Lab, and discovered the 21-cm line of hydrogen. Purcell received the Nobel Prize for this work.

Using borrowed equipment, the Dutch astronomers confirmed the discovery. Then they used the 21-cm line to map the clouds in our galaxy. They showed that the Milky Way has spiral arms, like the Andromeda Nebula, and determined their shapes and rotation speeds. They also discovered that the gas clouds close to the galactic center were moving outward at speeds as high as 135 km/s as well as rotating rapidly. Oort developed a theory of differential galactic rotation on the basis of these observations that went a long way toward explaining the origin of spiral arms and a flattened disk.

COSMOLOGY GROWS UP

For much of the period between the world wars, cosmology was largely a theoretician's game. Albert Einstein's general theory of relativity provided the mathematical tools for the sport. Alexander Friedmann, a brilliant Russian, applied the theory in 1922 to show that at least three models of the universe were possible, depending on the mean density of matter. Starting from zero, the radius of curvature of space could increase without limit, or expand and then contract back to zero, or oscillate in size. Hubble's later discovery of the recession of the galaxies seemed to favor the first model. The age of the universe (estimated from the oldest stars) and the fact that it is expanding were for a time the only solid observational facts in the science of cosmology (but see note 3.5 on Olbers's Paradox).

Then, in 1960, Martin Ryle added one more. Ryle wondered how the radio galaxies might be distributed in space. If they were all equal in power and if they were uniformly distributed in space, then the number he observed at a given brightness should decrease as a power of the brightness, with an exponent of −1.5. When he plotted his data in this fashion (fig. 3.2) he found a

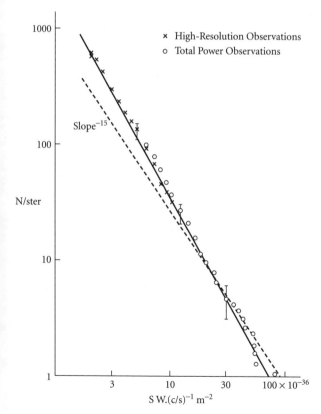

Fig. 3.2. Martin Ryle used his counts of radio sources to test the steady-state theory of Gold and Hoyle. Faint sources are more numerous than predicted.

marked deviation from his simple assumptions. That suggested two possibilities. Either distant galaxies were more numerous (per cubic light-year, say) than nearer galaxies, or they were brighter.

Ryle knew that the greater the distance of a galaxy, the longer it took for its light to reach Earth, and therefore the earlier the time when the light left the galaxy. When we look at distant galaxies, we look back in time. Therefore, both of his two possible alternatives led to the same conclusion: *the universe must be evolving in time.*

In the 1960s a great controversy was raging on the nature and origins of the universe. Hermann Bondi, Thomas Gold, and Fred Hoyle, three imaginative British astrophysicists, had proposed the *steady-state* model. The universe had neither a beginning nor an end, they said. It would always look the same, despite the fact that it is expanding. They proposed that, to fill the voids opened by the expansion, hydrogen atoms are created spontaneously from some source of energy in intergalactic space. In contrast, George Gamow, a Russian-born physicist at George Washington University, came to the idea that the universe was born in a highly compressed, extremely hot condition and had been expanding and cooling ever since (note 3.6). Hoyle dismissed Gamow's model as a "Big Bang" and fiercely defended his own model.

Ryle's conclusion, that the universe is evolving, argued against the steady-state theory, but a more convincing test was needed. Gamow pointed out that if he was right, space should be filled with the remnants of the original fireball, which, he estimated, had cooled to about 10 degrees kelvin above absolute zero. Robert Dicke, a physicist at Princeton University, recognized that at such a low temperature, the remnant would radiate at wavelengths of a few centimeters. The whole sky should glow at such wavelengths.

Dicke, an expert in microwave technology, set his students to building equipment to search for the glow, but they were scooped. In 1965, before they finished their work, two engineers at Bell Labs announced that they had discovered a strange glow over the sky that they could not account for. Dicke and his students met with Arno Penzias and Robert Wilson and explained to them that they had probably shot down the steady-state model and confirmed Gamow's Big Bang model.

With the discovery of this *cosmic microwave background,* astronomers now struggled to understand the physics of an expanding and cooling universe. Would it expand forever, or would it eventually collapse on itself? Were there any signs that the expansion was slowing down at great distances? What was the universe like in the first few millennia after the Big Bang? How could galaxies condense out of such an expanding broth of hot particles?

A STRANGE MANNER OF BEAST

Radio astronomy got a slow start in the United States, but the pace quickened after 1955, when John Bolton and Gordon Stanley joined the California Institute of Technology. Bolton and Stanley built a two-telescope interferometer in the Owens Valley, northeast of Los Angeles. Working at wavelengths as short as 30 cm and with adjustable baselines as long as 500 meters this instrument could locate a source to within 10 arcseconds, high precision for those days.

Bolton started a program to pinpoint radio sources and measure their sizes. Tom Matthews, a recent Harvard graduate from Canada, was hired to help do the work. In time he noticed that several sources were both very small and very bright. Among these was the source 3C48 in the constellation Triangulum. (The prefix 3C refers to Ryle's third Cambridge catalog.) What could these sources be?

Matthews determined the position of 3C48 within 10 arcseconds and persuaded Allan Sandage, a senior astronomer at the Mount Wilson and Palomar observatories, to photograph the patch of sky around it. In September 1960 Sandage did just that at the 200-inch telescope. The photographic plate showed a faint blue star with a tiny nebulosity extending from it. Nothing else in the error box around the center of 3C48 looked promising as a candidate.

In the following month Sandage returned to the 200-inch to take some spectrograms of this radio star. The spectra were very strange. They were marked with four very broad emission lines, which none of the distinguished spectroscopists at Caltech could identify with any known element. Sandage had also measured the optical brightness and color of this object and discovered that they varied between October and November. That argued for a very compact source, probably some kind of star (note 3.7). Sandage and Matthews reported their inconclusive results at a general astronomical meeting in December 1960, and there the matter rested for a while.

Meanwhile, at Jodrell Bank, Cyril Hazard had invented a clever technique. Some sources, he knew, lie on the path of the Moon around the sky. He could wait for the Moon to pass between the source and the Earth. At the instant when the Moon occulted the source, he could fix the time precisely. An exact time would determine the source's position in the east-west direction within a few arcseconds. In 1962 Hazard decided to try out his lunar occultation method on a source labeled 3C273 in the Cambridge catalog. Since he happened to be in Sydney, Australia, he asked for time on the new 64-meter steerable dish at Parkes, New South Wales. Despite some last-minute difficulties, he was able to pinpoint the source.

Back at Caltech, Maarten Schmidt heard about Hazard's result and thought he would see what he could find on the photographic plates of the Palomar all-sky survey. Like Sandage, he found a faint blue star at the source's location. And like Sandage, he decided to obtain a spectrum of the object with the Palomar 200-inch telescope. The spectrum was a real puzzle, a pattern of bright emission lines unlike any he had seen before.

After much cogitation, Schmidt understood why. He was looking at a familiar set of spectral lines of hydrogen, but in this object these lines had been redshifted in wavelength by 100 nm, or 15 percent (fig. 3.3). If the redshift was due to the Doppler effect, it would mean the object was receding at almost one-sixth the speed of light (45,000 km/s). And if that recession was due to the expansion of the universe, it would mean that this *quasi-stellar object* was at a distance of 3 billion light-years. How could such a distant object be visible? Even crude estimates of the power it must emit were staggering. Schmidt announced his result to a stunned scientific community, and the hunt was on for more of these quasars. Looking back, we can see that Schmidt was lucky. It turns out that only 1 percent of quasars emit powerful radio waves.

At first most scientists were unwilling to concede that quasars are very distant objects, because the energy requirements were so extreme. But in time even more extreme examples were found, with recession velocities up to 90 percent of the velocity of light (note 3.8). It was inconceivable that such fast objects could long remain near our galaxy. In addition, some of these quasars were found to vary in brightness over the course of only a few hours, which implied that their energy source is extremely compact.

A viable energy source had to be found. Some scientists thought that new laws of physics were involved, and in a sense they were correct. After all other possibilities had been eliminated, there remained only gravitational energy. At the center of each of these distant objects there had to exist an enormous, highly concentrated mass that sucked in the surrounding stars and gas and

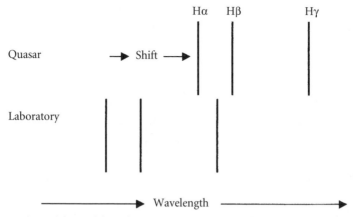

Fig. 3.3. Hydrogen lines (H alpha, H beta, and H gamma) are redshifted by 15 percent in the spectrum of the first quasar, 3C273 *(top)*. A laboratory spectrum *(bottom)* was used to calibrate the wavelength scale.

converted their gravitational energy to radiation. This central mass was postulated to be a supermassive black hole.

BLACK HOLES

John Wheeler, a physics professor at Princeton University, coined the term *black hole* to describe an object that had until then existed only in the minds of few theorists. In 1916 Karl Schwarzschild solved Einstein's equations of general relativity in the vicinity of a pointlike mass and showed how not even light could escape from its enormous gravitational field. But how could such a strange object form?

The Indian-born astrophysicist Subramanyan Chandrasekhar took the first step in 1930. He asked about the fate of a star that has exhausted all its nuclear fuel and begins to collapse under its own gravity. He proved that only stars with a mass less than a critical limit (1.4 solar masses) can collapse into a stable white dwarf. More massive stars must collapse even further.

Then in 1939 J. Robert Oppenheimer and H. Snyder proved that only stars with less than *three* solar masses could collapse to form a stable neutron star. (In a *neutron star,* matter is packed as tightly as the nucleus of an atom. A typical diameter is about 20 km.) Stars with masses greater than about three solar masses could collapse indefinitely to form a pointlike object, a black hole.

Once a black hole is born, it can feed on all the stars and gas in its vicinity, growing perhaps to a mass of hundreds of millions of Sun-like stars. This was the monster that was proposed to explain the enormous power output of quasars. A hole lies at the center of an otherwise normal galaxy and converts the gravitational energy of in-falling stars to radiation. This was a plausible explanation, but until a black hole, even a stellar-mass black hole, could be found, it remained just the best hypothesis. And the details of how radiation could possibly escape from a black hole remained unclear.

In the following years optical astronomers discovered distant galaxies that were not radio sources but emitted enormous fluxes of blue light from a small central nucleus. In fact, the nucleus emitted more light than all the rest of the galaxy. These *active galactic nuclei* often flickered rapidly, indicating that they could be no larger than a few light-hours in size. Once again, black holes were invoked to provide a powerful source of energy within a small space.

With the discovery of quasars and active galactic nuclei, the physics of black holes became a hot topic. Many important theoretical discoveries were made in the 1970s by such luminaries as Stephen Hawking, Roger Penrose, and Roy Kerr. Black holes, it seems, are not quite geometrical points of mass *(singularities)* but have a finite volume. They also rotate, and in-falling gas can accumulate in a flat, spinning *accretion disk* perpendicular to the rotation axis. A powerful magnetic field may be generated by the spinning mass. And so on.

Such theoretical results were fascinating, but as late as 1979 Sir Martin Rees, an expert in the field, surveyed the observational evidence for black holes and concluded that "although there are strong reasons for suspecting that some astronomical systems contain black holes, their existence has not yet been confirmed" (*Royal Society of London Proceedings* A368, no. 1732, pp. 27–32, 1979). Many astronomers were not so skeptical, however. They had been persuaded, in part, by the discovery of a strange type of radio galaxy, the double source.

JETS AND LOBES

In the early 1960s astronomers began to make crude images of extragalactic radio sources using interferometers. They soon learned that some sources in the Cambridge 3C catalog are extremely broad and consist of two well-separated patches (or *lobes*) of emission that lie on opposite sides of a host galaxy. The lobes were presumably cylindrical objects, giving the whole structure the shape of a dumbbell.

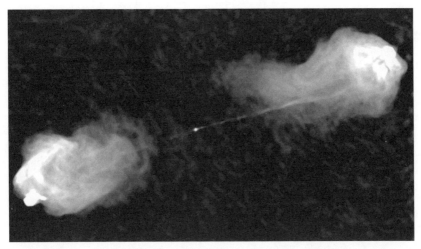

Fig. 3.4. Cygnus A is an active galaxy. In this 6-cm Very Large Array image, jets of relativistic electrons extend tens of thousands of light-years out from the nucleus of the galaxy and collide with the intergalactic gas to form two huge lobes. The lobes are separated by about 500,000 light-years.

The classic case of a double radio source is Cygnus A (fig. 3.4). In 1951 Baade and Minkowski had identified the source as a peculiar galaxy. From its red-shifted spectrum they determined that it lies at a distance of 1 billion light-years. That meant that the lobes were separated by nearly 500,000 light-years, making it one of the largest single objects known. (Actually, 3C236 is even larger, 20 million light-years across.)

As radio interferometers improved in resolution, the finer details of these strange double sources were revealed. Ryle's Five-Kilometer Telescope, the Westerbork array in Holland, and the three-element array at Greenbank, West Virginia, were soon producing images with an angular resolution of 10 arc-seconds or better. They were discovering narrow *jets* that extended a few thousand light-years from one lobe toward the other. Jets like these had been seen earlier in photographs of such galaxies as Messier 87, but there had been no indication that they might emit radio waves.

Cygnus A contains one of the most stunning examples of a jet. The image in figure 3.4, obtained with the Very Large Array in New Mexico in the late 1970s, shows the whole source at a resolution of about 1 arcsecond. Two narrow jets extend all the way from a tiny nucleus in the galaxy to two bright "hot spots" in the lobes. The jets contain bright little blobs that stream toward the lobes at supersonic speeds. Evidently the jets carry some form of energy from the galaxy far out to the lobes. Because the jets and the lobes are visible in *synchrotron radiation* (note 3.9), they must contain high-energy electrons and magnetic fields. Are these jets, then, a kind of electron beam? If so, how are they produced in the core of the galaxy? How do the jets preserve their shape over such huge distances?

Additional images of Cygnus A obtained with the Very Long Baseline Array (note 3.10) show the central galaxy and bits of the jets at a resolution of several *thousandths* of an arcsecond. Those images demonstrate that indeed the "engine" at the core of the galaxy is exceedingly small, perhaps only as large as our solar system.

The best guess is that this engine is a supermassive black hole, with a mass of a hundred million Suns. As stars, gas, and dust are sucked into the hole, they are thought to spiral into a flat accretion disk centered on the hole. As the disk rotates, it winds up its magnetic fields. This process accelerates electrons to relativistic energy and ejects them somehow in two collimated jets. The jets presumably interact with the surrounding intergalactic gas to produce the synchrotron radiation of the lobes.

This is about the best scenario offered to explain these bizarre objects, but the details are still being worked out. Nevertheless, Cygnus A and double sources like it emphasized the importance of nonthermal physical processes in the violent universe and lent additional support to the existence of supermassive black holes.

LITTLE GREEN MEN?

Antony Hewish, Ryle's colleague at Cambridge, had a clever idea for finding quasars. From data collected during the flight of *Mariner 2* in 1962 he learned that the Sun emits a wind of ionized gas (a *plasma*) that blows throughout the solar system. Hewish guessed that a pointlike quasar, observed through this plasma, would twinkle at radio frequencies for the same reason a star twinkles as seen through the Earth's atmosphere. A normal galaxy, with a finite width, would not. To test the idea he had his students build a cheap array of antennas covering 5 acres.

Jocelyn Bell was one of these students, and her task, after the array started working in 1967, was to search the hundreds of yards of chart recordings for characteristic twinkles. After a few weeks she found a signal consisting of a series of pulses 1.3 seconds apart. Hewish dismissed her find as manmade interference, but the source of the signal moved across the sky at the same rate as the stars. Could this be a signal from "little green men" in a distant galaxy? Bell was persistent and continued to search. To her great surprise and joy, she found another and then another and eventually a definite four. They all had periods of around 1 second.

It was highly unlikely that four different alien civilizations were signaling Earth. These sources had to have a more logical explanation. In 1968 Thomas Gold, the famous colleague of the famous Fred Hoyle, suggested that these *pulsars* were rotating neutron stars. Their powerful gravitational fields pull electrons and ions into the magnetic funnels at their poles. The electrons spiral around the magnetic field and emit synchrotron radiation in a highly focused beam. If the star were oriented so that the beam sweeps past the Earth, Hewish's array would pick up a periodic signal.

Here was the first evidence that these exotic creations, neutron stars, actually existed. In 1969 another group of radio astronomers discovered a pulsar in the Crab Nebula, the well-known remnant of a supernova. This discovery was particularly important because it lent support to a proposed connection between supernovas, neutron stars, and synchrotron emission.

Back in 1934 Walter Baade and his colleague Fritz Zwicky had conjectured that when a massive star has burned all its nuclear fuel, it collapses and forms a tiny neutron star at its core. They also guessed that when the collapsing envelope rebounds as a supernova, it produces cosmic rays, those energetic electrons and nuclei that stream through the galaxy. The Crab Nebula was possible proof of this scheme. The Crab's pulsar was the proposed neutron star, and the radio emission from the nebula was evidence for the presence of energetic electrons.

The discovery of pulsars created a new branch of astrophysics devoted to their internal structure and magnetic fields. As we shall see in a moment, neutron stars in binaries produce x-rays, and the mechanism for this emission proved to have important applications in galaxies.

In 1974 Hewish shared the Nobel Prize in physics with Martin Ryle, with only a nod to Jocelyn Bell. In that same year Russell Hulse and Joseph Taylor discovered a pulsar as a member of a binary system. Tracking the system for several years at the 305-meter Arecibo radio telescope in Puerto Rico, they demonstrated that the binary was losing its orbital energy by radiating gravitational waves, in precise accord with general relativity (note 3.11). They too received a Nobel Prize for their work, in 1993.

STELLAR NURSERIES

How do the stars form? We know that some stars are young because they shine with luminosities hundreds or thousands of times as great as the Sun and can have done that for only a few million years. Such young stars are invariably associated with gas and dust clouds. The Orion Nebula is a classic example of such a *stellar nursery,* in which dozens of stars are being born even today.

Astronomers have always wanted to learn just how a gas cloud collapses into a star, but the dense clouds of dust obscure the details of the process. As we shall see shortly, the infrared light emitted by warmer dust reveals *where* a star is being born. The real breakthrough came, however, when radio astronomers began to survey the sky at wavelengths of centimeters, millimeters, and even submillimeters. They discovered to their amazement that the gas between the stars contains a rich brew of molecules.

Walter Adams, a spectroscopist at Mount Wilson Observatory, was among the first to discover interstellar molecules. In the 1930s and 1940s he detected the simple molecules CH and CN from the imprint of their absorption lines on the visible spectrum of distant hot stars. Then after a pause of two decades, Alan Barrett and his colleagues at MIT discovered the spectral line of the OH hydroxyl molecule at 18 cm, in the radio spectrum of Cassiopeia A, one of the strong radio sources. From the Doppler shifts of the line they could determine the speeds and numbers of gas clouds along the path to the source.

Simple diatomic molecules might have been expected to form in the cold reaches of space, but when interstellar *formaldehyde* (CH_2O) was discovered in 1969, astronomers had to reconsider. Such complex molecules could form only in extremely dense, cold clouds and only in the presence of dust grains. Temperatures as low as 10 kelvin and gas densities one hundred times larger than average would be needed.

There was more to come. Charles Townes, the inventor of the laser, discovered interstellar ammonia (NH_3) in 1968. Two years later Arno Penzias and Robert Wilson, working at Bell Labs, discovered carbon monoxide (CO) in the Orion Nebula, at a wavelength of 2.6 mm. That same year George Carruthers, at the Naval Research Laboratory, launched a rocket to observe the ultraviolet spectrum of a star, Xi Persei, and found the telltale spectrum of interstellar molecular hydrogen. It turned out that most of the hydrogen in the galaxy is in this form.

By 1974 radio astronomers had identified more than twenty interstellar molecules, primarily by their emission or absorption at millimeter wavelengths (note 3.12). Some of them had as many as seven atoms, which confirmed the existence of extremely cold and dense gas clouds. A cloud could contain as much mass as one million Suns and extend over tens of light-years. (The size of our solar system is only about 6 light-*hours*.) These *giant molecular clouds* provide the ideal conditions for stars to form, and indeed, many young stars and proto-stars were found within them.

Astronomers began to map the clumps and knots within the clouds at ever-increasing resolution and began to unravel the complicated chain of events that allow a cloud to cool and condense into a star. Star formation became one of the hottest subjects in astronomy.

A STRANGE AND WONDROUS LIGHT

Shortly after the end of World War II, American scientists began to use V2 rockets, captured from the Germans, to launch instruments into space. In the beginning they had no idea of the variety of weird objects they would discover once they could detect cosmic x-rays and gamma rays. (See figure 3.1 to locate these rays in the electromagnetic spectrum.)

As the brightest object in the sky, the Sun was naturally the first target. Richard Tousey and Herbert Friedman, physicists at the Naval Research Laboratory (NRL), were among the pioneers. In 1947 Tousey and his colleagues obtained the first ultraviolet spectra of the Sun, and in 1949 Friedman detected the first solar x-rays. These discoveries confirmed that the Sun, a normal garden-variety star, possesses a million-degree corona. Other stars were expected to shine in x-rays, but as it turned out, stellar coronae are relatively weak sources.

Almost ten years passed before x-ray detectors became sensitive enough to search for cosmic sources (see note 3.13 about detectors). Then in 1962, Riccardo Giaconni and his associates at the American Science and Engineering Company decided to search for solar x-rays reflected by the Moon. They launched three Geiger counters on an Aerobee rocket and struck gold. Instead of the Moon, they discovered the first discrete cosmic x-ray source, Scorpius X-1, as well as a diffuse x-ray background coming from all directions. Photons as energetic as 10 kiloelectron-volts (keV) were coming from the sky. This was the true beginning of x-ray astronomy.

A flurry of discoveries followed quickly. For example, in 1963 Stuart Bowyer and friends at NRL found ten discrete sources near the galactic plane, including the center of the galaxy, the Crab Nebula, and several other supernova remnants. Most important, they showed that both the x-rays and the optical light from the Crab Nebula were most likely produced by the synchrotron emission of relativistic electrons, a process that was known to produce the Crab's radio emission. This was further evidence that supernovas produced high-energy particles, as Baade and Zwicky had proposed in 1934. But exactly how does this happen, and how are the magnetic fields generated that synchrotron emission also requires?

Later the NRL group discovered that M87 (a giant elliptic galaxy) and the Large Magellanic Cloud (the nearest galaxy to our own) were emitting x-rays. And a group at the Goddard Space Flight Center discovered x-ray emission from a 100-million-degree cloud of gas that lies *between* the galaxies of the Coma cluster. X-rays seemed to come from everywhere.

The biggest puzzle was the nature of the strong discrete sources, such as Scorpio X-1. In 1966 a fairly accurate location for this source was relayed to Alan Sandage at the Palomar 200-inch telescope. He found a faint blue star in the error box, and its spectrum convinced him that it was an old nova. Such novas were known to arise in binary systems, but no companion star was visible. Several theoreticians suggested immediately that the x-rays are produced by gas that heats as it falls onto a compact companion star, a white dwarf or a neutron star. But a test of this idea would have to wait for more data.

In 1970 x-ray astronomy took a giant step forward with the launch of Uhuru (note 3.14), the first satellite devoted entirely to x-ray astronomy. Riccardo Giaconni and his team at American Science and Engineering had equipped the satellite with two x-ray instruments pointing in opposite directions. Although these were not true imaging telescopes, they did yield sufficiently accurate positions to enable optical astronomers to identify many of the sources. In a flight of three years, Uhuru discovered over three hundred sources all over the sky, some of which were identified with supernova remnants, peculiar galaxies, and clusters of galaxies.

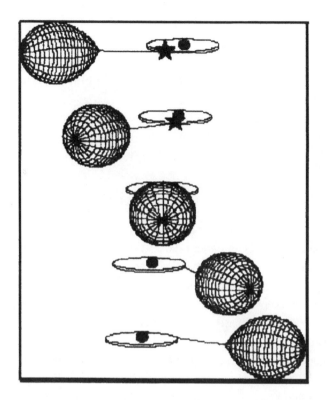

Fig. 3.5. A sketch of an x-ray-eclipsing binary. A neutron star pulls gas off its larger companion and into an accretion disk.

And, most interesting, Uhuru discovered a number of sources, like Hercules X-1 and Centaurus X-3, that pulsate in brightness with periods of around a second. They also disappear and then reappear after about a day. A simple model was soon proposed in which a neutron star and a normal star orbit around a common center as a binary system. Periodically the neutron star swings behind the larger star and is eclipsed (fig. 3.5).

But how does the neutron star generate x-rays? A physical model was proposed in which a neutron star drags gas off its companion with its powerful gravity. As the gas falls inward, it heats to tens of millions of degrees, to the point where it radiates x-rays. This conversion of gravitational to radiative energy was calculated to be extremely efficient. Here was a mechanism that could explain the enormous emission of such objects as quasars and galaxies with abnormally bright centers. It was a discovery of the first rank.

There was a small problem with this kind of model, however. In-falling gas would have too much spin to be captured directly by a neutron star. Instead, as K. Prendergast and G. Burbidge suggested in 1968, the gas would spiral into an accretion disk surrounding the neutron star. Friction within the disk would slowly feed mass onto the star's surface, where the x-rays were emitted. Such a model accounted nicely for the behavior of Scorpio X-1, the brightest x-ray source in the sky.

Accretion disks were recognized as important factors in interacting systems of various kinds. They could, for example, appear as a star is formed from the interstellar medium.

STELLAR-MASS BLACK HOLES

Eclipsing binaries are especially valuable because they provide sufficient data to determine the masses of their component stars. In most x-ray binaries, the mass of the large normal star was found to be 10 to 20 solar masses, and that of the neutron star about 1 or 2 solar masses.

Cygnus X-1 is a wild exception. It evidently consists of a blue star with a mass of about sixteen Suns, and a neutron star with a mass of about nine Suns. But hold on! Oppenheimer and Snyder had proved that a neutron star cannot exceed a mass of about three Suns. A more massive star will collapse into a black hole. So here in Cygnus was the first candidate for a black hole with a mass comparable to a single heavy star. Later, other candidates were found, like V404 Cygni. But many details remained to be worked out. How, for example, can a black hole emit radiation? Does it form an accretion disk, like those surrounding neutron stars in binaries?

THE GAMMA-RAY SKY

Gamma rays are extremely hard x-rays, photons with energy greater than several hundred thousand electron volts. In the 1950s several theorists predicted that supernovas and colliding galaxies might emit gamma rays. Cosmic rays, the fast electrically charged particles flying around in space, might also produce them as they collided with neutral gas clouds. If these energetic photons could be detected, one might learn more about the most violent events in the universe.

However, cosmic gamma rays are absorbed high in the Earth's atmosphere. So astronomers had to wait until 1961 for their first glimpse of the gamma-ray

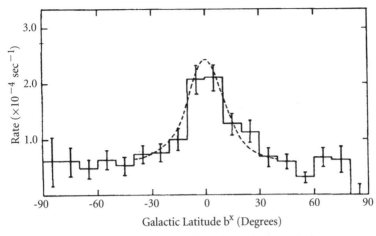

Fig. 3.6. George Clark et al. discovered that the plane of our galaxy emits high-energy gamma rays. The galactic center stands out as a strong source.

sky. In that year, NASA launched Explorer XI. In its short life this small satellite detected fewer than 100 gamma-ray photons, but they were distributed randomly over the sky. That at least suggested cosmic sources, possibly diffuse, possibly pointlike. Gamma-ray astronomy had been born.

The next major discoveries came seven years later. An experiment aboard the third Orbiting Solar Observatory was able to map a band in the sky at photon energies above 70 million electron volts. To everyone's surprise, the whole disk of our galaxy showed up clearly in the map, with broad peak intensity toward the galactic center (fig. 3.6). How could one account for such broad diffuse emission?

That same year, another group launched a high-altitude balloon and detected gamma rays from that source of all good things, the Crab Nebula. The spectrum, up to 500 keV, joined nicely to the radio and optical spectra and confirmed that at least in the Crab, gamma rays were produced as the synchrotron radiation of relativistic electrons trapped in magnetic fields. But supernovas are relatively rare events in our galaxy and couldn't account for the emission of the disk. The question continued to puzzle the experts.

Cos B, launched in 1975 by the European Space Agency, was the first satellite to map the entire sky with enough resolution to begin to locate discrete sources. By 1980 it had discovered twenty-five of them, all concentrated in the plane of the galaxy, emitting photons with energy above 100 million electron-volts (MeV). Four could be identified with supernova remnants. The rest remained a mystery.

In the following years some pulsars and x-ray binaries in our galaxy were found to emit gamma rays, either periodically or in spurts. But the source of the diffuse emission of the galaxy remained uncertain. Then yet another discovery startled the observers.

BURSTERS

During the 1960s the U.S. Defense Department flew pairs of Vela satellites to monitor the compliance of the Soviet Union with the terms of a treaty banning the test of nuclear weapons in space. The satellites were equipped with x-ray and gamma-ray detectors with fast time responses. Although the satellites never caught the Soviets misbehaving, they did record the first cosmic gamma-ray burst. This was a blast of photons with energy up to 1.5 MeV lasting only a few seconds.

In three years the Vela satellites detected sixteen bursts of gamma rays, each lasting only a few seconds. Although initially the experimenters couldn't fix the positions of these bursts, they could exclude the Sun and the Earth as sources and suggested that the bursts originated outside our galaxy. The strength of the bursts was consistent with the amount of gamma-ray energy a supernova was predicted to emit from a distance as great as 3 million light-years.

For over two decades experimenters continued to detect these brief bursts but without sufficient positional information to track down the sources. Astrophysicists divided into camps, those who accepted a cosmic origin and those who preferred an origin in our galaxy. The essential question was how a very distant event might produce such quantities of gamma rays. A fierce debate was carried on in the journals. Only with the flight of the Italian-Dutch gamma-ray satellite BeppoSAX and the subsequent detection of an optical afterglow in 1997 was the issue settled in favor of a cosmic origin. The actual cause of a gamma-ray burst is still controversial.

FAINT HEAT FROM A COLD SKY

In a completely dark room, your face emits enough heat for a sensitive detector to make a recognizable image. Perhaps you've seen a demonstration of such night-vision devices at a science fair. This invisible heat radiation lies in the *infrared* part of the spectrum, at wavelengths between 1 and about 300 microns. As figure 3.1 shows, infrared light leaks through the atmosphere in only a few windows. But Earth-bound astronomers learned to make good use of what they could capture.

Infrared light offers the astronomer several advantages. For example, the center of our galaxy is hidden from us by thick interstellar dust clouds, which absorb and scatter visible light. Infrared light, however, is absorbed far less and permits astronomers to peer into the heart of the galaxy. And as we've seen, distant objects like quasars are redshifted and appear most favorably at infrared wavelengths. Finally, some of the coldest objects in space, such as interstellar gas and dust, emit most of their radiation in the infrared portion of the electromagnetic spectrum (note 3.15) and are best studied in such light.

Despite these advantages, infrared research developed rather slowly. Astronomers had to wait until detectors with high sensitivity at long wavelengths became available (note 3.16). In the 1920s, thermocouples were the state of the art, but they could do no more than measure the temperatures of the planets and a few bright stars. Lead sulphide cells, in common use by 1950, were a great improvement. When cooled in liquid nitrogen to reduce the signal from a warm telescope, they were sensitive to wavelengths as long as 4 microns. During the mid-1960s Robert Leighton and Gerry Neugebauer, professors at Caltech, used them to undertake the first comprehensive survey of the northern sky at wavelengths near 2 microns. During the course of the survey, the Orion Nebula was found to be a strong infrared source.

Neugebauer decided to look further. In 1965 he and Eric Becklin made more detailed observations at 2 microns, using the 60-inch (1.5 meter) telescope at Mount Wilson. They discovered seven point sources. Of these, six could be identified with visible stars, but the seventh was a mystery. So they went to the 200-inch telescope and made new measurements at four wavelengths. From these they could determine that the object had a bright central core and extended wings of emission. Its temperature was estimated at about 700 kelvin.

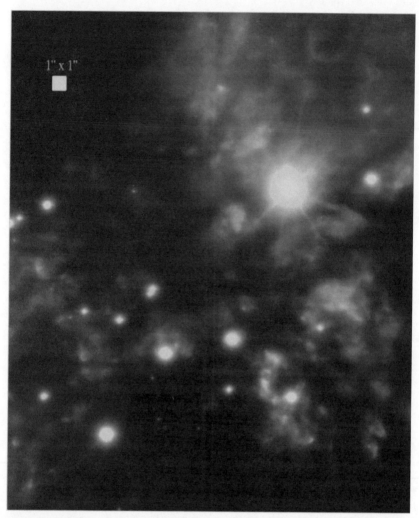

1" x 1"

Fig. 3.7. The Becklin-Neugebauer cloud (the bright starlike object surrounded by wispy clouds) and the Kleinmann-Low nebula lie within the Orion Nebula. Shock-excited molecular hydrogen emits a line at 2.12 microns (rendered blue in the image; see color gallery), and ionized hydrogen emits a band around 2.2 microns (rendered in red).

Becklin and Neugebauer concluded that the object was probably a dusty hydrogen cloud in the process of forming a new star (fig. 3.7).

Within a year, D. E. Kleinmann (Rice University) and Frank Low (University of Arizona) had announced the discovery of another star being born. They too had observed the Orion Nebula, but with a new type of detector, the *germanium bolometer*. When immersed in liquid helium at 4 kelvin, this detector was sensitive to wavelengths as long as 28 microns. Kleinmann and Low found an extended object not far from the Becklin-Neugebauer source emitting strongly at 22 microns. This source was even colder, with an estimated temperature of not more than 150 kelvin. The two astronomers concluded that "the discovery of the infrared nebula in Orion means that the early stages of star and star-cluster formation can be observed in the far infrared" (*Astrophysical Journal* 149, L1, 1967). A new era in stellar astronomy had begun.

In the following decade the physics of star formation was studied intensively. We learned that interstellar dust plays an essential role in radiating the heat that the collapse of a hydrogen cloud releases. Without such efficient cooling mechanisms, a star could not form by means of gravitational contraction. Second, we learned that stars form preferentially in the densest parts of hydrogen clouds, where the temperature is cool enough for molecules to form. (Recall how the radio astronomers have been finding such clouds.) Third, we learned that young stars are often surrounded by flat, spinning accretion disks that absorb the angular momentum of the contracting gas. (These disks are thought to be the forerunners of planetary systems, and much current research is devoted to exploring such disks.) Finally, stars in the throes of birth were seen to eject winds or jets of gas, for reasons that were unclear at the time.

Astronomers used a variety of tools to explore the infrared sky. Gerard Kuiper, an eminent planetary astronomer, established the Mount Lemmon Observatory near Tucson, Arizona. In the arid conditions there, Kuiper made pioneering studies of the planets in the infrared with his 1.5-meter telescope. Then in 1974 NASA named a flying observatory for him. The Kuiper Airborne Observatory is a converted jet transport that carries a 0.92-meter telescope and cooled detectors high into the stratosphere, where wavelengths as long as 500 microns are accessible. This national facility soon began to produce a flood of new science.

High-altitude balloons were also used in the mid-1970s, especially to test the Big Bang theory. They carried instruments to scan the sky at submillimeter wavelengths in order to obtain the spectrum of the cosmic fireball.

But the ultimate platform to study the infrared sky was an orbiting satellite. Planning for the Infrared Astronomical Satellite, a joint project of the United States, the United Kingdom, and the Netherlands, began in the 1970s, and the bird was launched in 1983. We'll survey some of its findings in a later chapter.

TWO DISCOVERIES IN VISIBLE LIGHT

The 1960s and 1970s were fabulous decades in astronomy. Radio and x-ray astronomers seemed to announce the discovery of an exotic object almost every month. Orbiting satellites were being launched at a furious pace, and interplanetary probes were cruising the solar system. In the midst of all this excitement, optical astronomers were also making some surprising discoveries. Among the most telling was a little matter of too little matter.

Fritz Zwicky first pointed out the problem in 1933. Zwicky was a Swiss-born astronomer and a professor at Caltech, considered an eccentric by his colleagues. He in turn referred to them as "spherical bastards," meaning that they were bastards whichever way he looked at them. Gruff, opinionated, and brilliant, he suffered fools not at all. He had wide interests, including the origin of cosmic rays, the plasticity of crystals, the meaning of the recession of the galaxies, and the physics of supernovas.

In 1933 Zwicky was studying the giant cluster of galaxies in the constellation Coma Berenices. First he estimated the mass of the cluster by multiplying the luminosity of each galaxy by a well-determined factor and summing the masses. Then he derived a second estimate from the spread, about the mean, of the velocities of the galaxies. He was amused to discover that this second estimate was *four hundred* times larger than the first. Now, in astrophysics, discrepancies of a factor of two are not uncommon and are generally no cause for concern. But clearly this mismatch was telling him something. The cluster would fly apart if its mass were as low as the first estimate indicated. More mass had to be present, but whatever it was, it was invisible. What could it be?

No one pursued Zwicky's discovery until 1970, when Vera Rubin and Kent Ford Jr., astronomers at the Carnegie Institution of Washington, determined the rotational velocities of ionized hydrogen clouds in the Andromeda Neb-

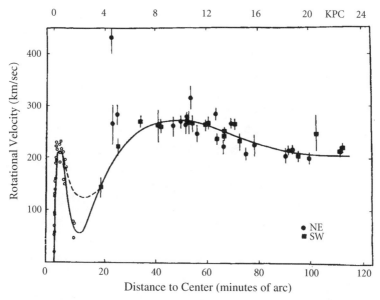

Fig. 3.8. From their study of galactic rotation, Vera Rubin and Kent Ford Jr. discovered that the Andromeda Nebula contains a considerable amount of invisible mass. The rotation curve flattens out far from the center, in contradiction to Kepler's third law.

ula, a spiral galaxy similar to our own. They expected, from Kepler's third law, to find that the velocities would decrease with increasing distance from the center, just as the velocities of the planets decrease with distance from the Sun. But instead the rotation curve leveled off and remained flat (fig. 3.8). Later, Rubin and Ford concluded tentatively that much more mass is present in the outer portions of Messier 31 than is visible. The term *dark matter* entered the astronomical vocabulary.

Subsequent studies of other spirals have confirmed these results. Astronomers faced the uncomfortable conclusion that as much as 90 percent of all the mass in the universe was unaccounted for. They would continue to search for this elusive mass well into the 1990s.

GRAVITY'S LENS

In 1979 three astronomers at the Multi-Mirror Telescope in Arizona discovered a most curious pair of quasars. They were separated in the sky by a mere

5.7 arcseconds, they had very similar spectra and the same redshift, $z = 1.4$ (z is the fractional shift in wavelength of a source that is caused by its recession; $z = 1$ means that each wavelength has been doubled). The astronomers pointed out that if this was a chance coincidence it must be extremely rare. On the other hand, they suggested that it might be the first example of *gravitational lensing* of a distant object by a massive nearer object.

The deflection of starlight by the Sun, observed during the total eclipse of 1919, was the first successful test of Einstein's theory of relativity. Sir Arthur Eddington, a participant in the eclipse expedition, suggested soon afterward that a massive object might form an *image* of a more distant object by focusing the light with its gravitational field. Fritz Zwicky realized in 1937 that such images could allow an observer to estimate the mass of the foreground object. Several theorists developed the theory of lensing in the 1960s and 1970s. Then with the discovery of these twin quasars in 1979, Zwicky's proposal appeared to be feasible.

Gravitational lensing would become a valuable tool in searching for the missing mass of galaxies. But to apply this technique would require many hours of surveying the sky and then obtaining accurate spectra. Bigger telescopes would be needed more than ever.

WHAT WE LEARNED

The three decades between 1950 and 1980 changed and broadened the views of astronomers. At the same time a host of new questions came to the fore:

- The universe had originated in a Big Bang and was expanding, perhaps to collapse again. But how old is the universe, and how fast does it expand? Somehow, out of this primeval fireball, clusters of galaxies formed. How was this possible?

- Supernovas were seen to play an essential role in the chemical history of galaxies. Heavy elements are cooked up in their innards and then spewed out to enrich the interstellar medium in a constant cycle of stellar death and rebirth. How does this cycle work, in detail?

- Neutron stars and black holes really do exist outside the minds of theorists. How do we explain their bizarre physical properties? Black holes with masses of hundreds of millions of solar masses seem to populate the centers of many galaxies and generate incredible amounts of radiant energy. How are these supermassive holes created? What is their role in the evolution of galaxies? And speaking of evolution, how do we account for the variety of shapes of galaxies?

These are only a few of the issues that astronomers would tackle in the succeeding decades.

THE RISE OF THE GREAT CENTERS

BETWEEN 1955 AND 1975, six great clusters of telescopes were built around the world. Blessed with new technology and adequate funding, astronomers were able to build telescopes with mirrors as large as 4 meters, and they were determined to place them only in locations where the best observing conditions prevailed. To find such ideal conditions, they were prepared to reach out across the oceans, to travel to the very ends of the Earth, if necessary. Scores of potential sites were tested, on mountains, lakes, and islands. Eventually astronomers converged on a few of the best and built dozens of telescopes there.

What drove this boom in construction? As we saw in the last chapter, the advent of manned space flight and unmanned satellites generated tremendous interest in astronomy and space exploration. X-ray and ultraviolet telescopes in space, and infrared telescopes on the ground, had discovered new worlds to explore. At the same time, radio astronomers had opened a new window on the universe and found a zoo of strange and violent objects. Optical astronomers were eager to follow up on these discoveries and to settle longstanding issues. To see fainter objects and make more precise measurements, they needed larger telescopes with more sensitive instruments. They banded together to put pressure on their governments to provide them.

In Europe the postwar era ushered in a new spirit of cooperation. Previously each nation had built its own optical observatories in the best locations available within its borders. For example, the French had major observatories at Paris, Meudon, Pic-du-Midi, and Haute-Provence. The Russians had built an observatory at St. Petersburg, the Germans at Berlin and Potsdam, the

British at Greenwich, and the Dutch at Leiden, Groningen, and Utrecht. Despite their generally wet and cloudy climate, Europeans had made most of the great discoveries in astronomy using telescope mirrors no larger than 2 meters. Later, with the completion of the 200-inch (5-meter) Palomar telescope in 1948, the Americans took the lead in observational cosmology. European astronomers were keen to compete. In the late 1950s they were ready to join forces to build a major observatory in a favorable climate.

In the United States, the opening of the Space Age ignited a huge interest in astronomy among the public. Schools of astronomy and space science were sprouting everywhere, and an army of newly minted astronomers wanted access to world-class optical telescopes. Unless they had contacts among the astronomers at Mount Wilson or Lick Observatory, however, they had few opportunities for groundbreaking research.

In time, several distinguished astronomers petitioned their congressmen and made the case that the United States needed a national observatory open to all qualified astronomers if it were to maintain its position as a leader in astronomy. They were successful, and in 1955 the newly established National Science Foundation decided to provide an optical observatory as one of its first initiatives. Soon afterward it agreed to establish the National Radio Astronomy Observatory.

AN OBSERVATORY IN A DESERT

The National Science Foundation has no charter to operate a facility like a national observatory. It can, however, delegate its responsibility to a qualified nonprofit organization. So in 1957 six American universities with strong astronomy departments (Harvard, Indiana, Ohio State, the University of Chicago, the University of Michigan, and the University of Wisconsin) banded together to form the Association of Universities for Research in Astronomy (AURA). This outfit would receive a charter from the NSF to build and operate an optical observatory, somewhere in the United States, to serve the needs of all American astronomers (note 4.1). The observatory was to have a capable staff, state-of-the-art instruments, and the most advanced telescopes. But first a site had to be found.

The criteria for an ideal site were well known and agreed upon. First, one wanted a dark sky, free of dust, smog, and light pollution. That immediately suggested someplace far from a major city and probably atop a mountain. But at the same time one wanted to be *near* a city, to allow easy access for construction and observing. One can't have both remoteness and proximity, so the astronomers had to compromise. They also had to think of the long-term future. When George Ellery Hale established Mount Wilson Observatory, Los Angeles was a sleepy little town surrounded by orchards. By the mid-1950s it had grown like a weed, and its bright lights were already a problem for the observatory. Similarly, the Lick Observatory, perched on Mount Hamilton above the booming city of San Jose, eventually lost its pristine skies. Even the Palomar Observatory, near San Diego, would in time experience a deterioration of its skies.

Second in importance to a dark sky is a generally cloudless climate. Astronomers everywhere, especially in Europe but also in the eastern United States, had fought cloudy weather for all of their careers. So where are the skies "not cloudy all day"? In the Far West, particularly in the desert Southwest. That would include southern California and parts of Arizona and New Mexico, a vast territory in which to search for an ideal site.

After a dark sky and few clouds, good *seeing* (sharpness of image, as determined by atmospheric turbulence) is probably next on the astronomer's list of priorities. Hale was originally drawn to Mount Wilson not only by its generally cloudless climate but by its excellent seeing. Where can one find a place where the stars don't twinkle, where the image of a planet is rock-steady? Of all the characteristics of a good location, seeing is the most variable and the most difficult to guarantee. Mountain sites have often been preferred, but usually for their lack of dust and haze rather than their seeing. Some mountains offer good seeing at least part of the year and part of the night, but until recently, predicting where such mountains are to be found has been a black art.

Finally, a site for a national observatory must have sufficient space for future growth, and the costs of providing roads, power, and water must be reasonable.

In 1955 AURA set out to find the best site for a national observatory. Helmut Abt and Aden Meinel, two young astronomers at the University of Texas's McDonald Observatory, agreed to take on the job. They were looking for a mountain somewhere in the Southwest. It should be higher than 2,000 meters, to avoid the hot turbulent air near the desert floor, but no higher than 2,600 meters, to avoid high winds and, possibly, snow. And its summit should be forested, to suppress local air turbulence.

Abt first studied topographic maps to get an overall impression. Then he chartered a small plane for a closer look. He was especially impressed with several mountains in the Papago Indian Reservation southwest of Tucson, Arizona. They were relatively isolated peaks and had a cover of small trees, which could help to reduce ground turbulence. Baboquivari Peak, a towering egg-shaped mountain was sacred to the Tohono O'odham Indians and hence off-limits, but Kitt Peak was a possibility. Abt and Meinel hiked up the mountain and camped overnight to find out whether the lights of Tucson were a problem. They were not (then); the sky was gorgeous.

Serious testing of Kitt Peak soon followed. In addition to the usual visual and photographic tests of seeing, a new technique was employed. Towers 20 meters tall were installed at several locations on the mountain. Temperature and wind sensors were mounted at several levels on each tower, and their data were recorded continuously. By comparing the temperature fluctuations at the different heights one could determine the amount of turbulence and its dependence on the local topography.

But Kitt Peak was only one of over 150 possible sites that had to be evaluated. After three years of testing, Kitt Peak was finally selected from a short list of eleven, as the best all-around candidate. Not only did it possess a dark sky, many cloudless nights, and good seeing, Kitt Peak was also close to the University of Arizona in Tucson, where many young astronomers were being trained. In 1958 the NSF signed a treaty with the Tohono O'odham for a lease on the mountain. Construction of a road and the design of the observatory buildings on the mountain and in Tucson began soon afterward.

The first nighttime telescope on the mountain was a 0.9-meter Cassegrain, followed soon by a 1.3-meter Cassegrain and a 2.1-meter Ritchey-Chrétien (note 4.2). In the early 1960s the McMath-Pierce Solar Telescope (fig. 4.1), with a 1.5-meter mirror, joined these other telescopes. This distinctive solar telescope is still the largest in the world.

Kitt Peak's crown jewel is the Mayall 4-meter (fig. 4.2), named for Nicholas Mayall, the observatory's first director. When the telescope was completed in 1973, it was second largest in the world. It still is a most impressive structure. It has a 15-ton mirror of fused quartz, the largest that could be cast with the available technology. The telescope is a conventional Cassegrain, with a focal ratio of f/2.7 at prime focus. (The *focal ratio* is the focal length of the mirror divided by its diameter. A small ratio offers a wide field of view and short photometric exposures.)

Fig. 4.1. The McMath-Pierce Solar Telescope is still the largest in the world, with a mirror 1.5 meters in diameter. Sunlight is directed to the mirror from a heliostat at the top of the tower, along a long air path.

The Mayall's equatorial mount features a horseshoe bearing for the polar axis that resembles that of the Palomar 200-inch. The dome has double walls to insulate the mirror from temperature swings. To preserve the fine nighttime seeing, the telescope and its 500-ton dome are perched on a tower 57 meters high, which can be seen from a distance of 80 km.

Kitt Peak has grown enormously since its founding. Presently there are *twenty-two* optical and radio telescopes on the mountain (fig. 4.3), including the 3.5-meter WIYN, which is shared by the University of Wisconsin, Indiana University, Yale University, and Kitt Peak's present parent organization, the National Optical Astronomy Observatory (NOAO). Astronomers from the United States and abroad compete for access to the telescopes by submitting detailed observing proposals to a standing committee. And the competition is often fierce. Not long ago five times as many nights were requested on the 4-meter as there are nights in a year.

Astronomers from all over the world have used the Mayall Telescope to make some important discoveries. In the early 1970s, for example, Vera Rubin, from the Carnegie Institution of Washington, found convincing evidence for the existence of dark matter when she measured the rotation profiles of nearby galaxies (note 4.3). And in 2001 Sangeeta Malhotra and James Rhoads, astronomers from Johns Hopkins University, discovered a dense concentration of energetic young galaxies at a redshift of $z = 4.5$. Why, they asked, are these young galaxies so bright? Is the reason a burst of star formation, supermassive black hole activity, or something peculiar in the evolution of a galaxy? Their study raised some basic questions on the formation of galaxies.

Fig. 4.2. The 4-meter Mayall Telescope on Kitt Peak

Fig. 4.3. Aerial view of Kitt Peak. The Mayall Telescope is in the tall white dome at the lower left.

THE SOUTHERN SKIES

By an accident of history and geography, most astronomers live in the Northern Hemisphere and are able to observe only the northern sky. The southern skies remained relatively unstudied, even as recently as the 1950s.

That's not to say that nothing was known. In the 1750s Abbé Nicolas Louis de Caille traveled to the Cape of Good Hope and compiled a catalog of the positions of ten thousand stars and some forty "nebulae." Among these was Omega Centauri, which was later recognized as the largest globular cluster in our galaxy. In the early 1820s James Dunlop, a skilled amateur in Australia, compiled a list of six hundred nebulae in the southern sky. And in the 1840s John Herschel (William's son) came to Cape Town to complete his all-sky General Catalog of deep-sky objects.

These pioneers were basically visitors to the South, however. The first permanent observatory below the equator was the Royal Observatory at the Cape of Good Hope, established in 1820.

In the 1890s David Gill, the director of the Cape Observatory, was engaged in a massive photographic survey of the southern sky. This was part of a joint effort with northern astronomers to determine the positions and brightness of the stars over the entire sky. Gill was desperately short of help in measuring the photographic plates, and when Jacobus Kapteyn, a young Dutch astronomer, volunteered to assist, Gill readily agreed. Their result, after thirteen years of labor, was a catalog of over four hundred thousand stars. Kapteyn went on to analyze these data and to derive from them the motion in space of the solar system. Eventually two other Dutch astronomers, Jan Oort and Bertil Lindblad, capped Kapteyn's discovery by determining the rotation of the Milky Way from these data.

Harvard University was another pioneer in the Southern Hemisphere. In 1865 Uriah Boyden, a wealthy Boston engineer, willed $228,000 to Harvard for a southern observatory. In 1891 Harvard used these funds to build a 0.6-meter refractor (with an Alvan Clark lens) on a mountain near Arequipa, Peru. Another photographic survey of the southern sky was completed, with the discovery of hundreds of nebulae, and thousands of variable stars. The spectra of thousands of stars were also obtained. Analysis of these spectra led to the modern system of classifying stars (note 4.4). In 1927 the observatory was transferred to the better climate of Bloemfontein, South Africa, and continued its work with a 1.5-meter.

KITT PEAK LOOKS SOUTH

Kitt Peak National Observatory (KPNO) began to plan in 1958 for a sister observatory in the Southern Hemisphere. A year later Jurgen Stock, an intrepid young astronomer, was dispatched to scout out the best observatory sites in South America. He was soon drawn to the arid climate of northern Chile. There, the foothills of the Andes Mountains overlook the Atacama Desert, one of the driest, least cloudy regions on Earth. He spent several years taking observations of clouds, dust, haze, wind, and temperature. He also compared the seeing at pairs of sites with a new type of instrument, a double-beam telescope (note 4.5).

Northern Chile has much to recommend it as a possible location for an observatory, but occasionally dust and haze can be a problem. Stock moved further south. He narrowed his search to two mountains in the foothills, Cerro Tololo and La Peineta, about 80 km east of the pleasant city of La Serena. He discovered that the seeing was excellent at both locations, but that Peineta had much higher winds during the winter. That eventually tipped the decision to Tololo.

One cannot simply buy a mountain in Chile, even if one wished to. Instead, KPNO and its parent organization AURA approached officials at the University of Santiago and proposed to share an observatory on Tololo. The university recognized the benefits of cooperation for its students and recommended the scheme to the Chilean government. In time a partnership was approved, and the Cerro Tololo Inter-American Observatory (CTIO) was created. Jurgen Stock was appointed interim director.

There followed a period of rapid construction. As you can see from figure 4.4, the top of Tololo had to be removed to provide sufficient flat space. The first telescopes to arrive were in the meter class, including the University of Michigan's 0.6-meter Curtis-Schmidt Telescope. In the early 1970s a 1.3-meter Cassegrain, a 1.5-meter Ritchey-Chrétien, and a 2.1-meter telescope were built. Then, in 1974, a twin of Kitt Peak's 4-meter was completed at Tololo and named for Victor Blanco, the first director.

Fig. 4.4. Cerro Tololo Inter-American Observatory. The large dome houses the Blanco Telescope.

Cerro Tololo has proved to have some of the best seeing anywhere in the world. In addition, its dry climate has made it a Mecca for research in the infrared. It has attracted astronomers from all over the world. The last time I was there we watched a giant condor feeding just outside the window of the observatory's cafeteria, a wonderful accompaniment to lunch.

The Blanco 4-meter has enabled astronomers to make many important discoveries. For example, a team from Harvard's Center for Astrophysics and Canada's National Research Council discovered three more moons of Neptune in January 2003. That brings the total to eleven. Even more exciting, a large team from the Lawrence Berkeley National Laboratory announced in 1998 that the expansion of the universe is accelerating. They based their findings on observations made with the Blanco telescope of over seventy-five supernovas at huge distances. We'll have much more to say about all this in a later chapter.

HERE COME THE EUROPEANS

In 1953, while Stock was searching in Chile for the ideal site, he was visited by a team of Europeans who were also looking for a place to build an observatory.

Walter Baade, the distinguished Dutch-born astronomer (note 4.6), was the first to propose a southern observatory that would be shared by all the astronomers of Europe. In 1954 he invited twelve astronomers from six nations (Belgium, France, Germany, Great Britain, the Netherlands, and Sweden) to meet in Leiden to discuss his idea. The time was ripe for action, and the astronomers had no trouble in reaching a consensus. They resolved find the funds to build a shared 3-meter telescope somewhere below the equator.

The first step was to approach their national scientific councils and begin to lobby for the project. Progress slowed markedly once the national governments became involved, as bureaucrats and politicians haggled over the terms of the agreement. How would the financial costs be assigned, and how would time on the telescopes be allocated? Great Britain left the consortium in 1960 to follow its own plans in Australia. (Eventually Britain rejoined the Europeans, however.) After years of negotiations, a formal convention was signed among the nations in 1962. The European Southern Observatory (ESO) was a reality.

During this whole period, a suitable site was sought in South Africa, where several nations had some limited observing experience. After eight years of testing, the ESO survey team gave up on South Africa because even at the best sites, the seeing was too variable. Meanwhile, they had learned of Stock's prospecting in Chile and decided to investigate.

In June 1963 an ESO team arrived to compare the virtues of Cerro Tololo and a nearby mountain on the CTIO reservation, Morado, which had a larger area near the summit and similar seeing conditions. They also tested three neighboring mountains. At first the ESO astronomers considered building at Morado, a plan that would enable them to share the costs of a road and other support facilities with the Americans. In 1964, however, they finally opted for La Silla, a mountain some 200 km north of Tololo in the Andean foothills.

In deciding to build an observatory in Chile, the ESO Council had to weigh the risk of a large earthquake. Chile lies at the edge of a tectonic plate and is prone to severe earthquakes. On May 22, 1960, for example, the largest earthquake of the twentieth century (Richter magnitude 9.5) devastated several cities. But such a monster would probably not recur for another century. The chance was worth taking.

After the usual struggle to build a road and provide power and water, La Silla began to operate as an observatory in 1965. The first telescopes were modest, but in November 1976 ESO proudly unveiled a telescope with a mirror 3.66 meters in diameter. Then in 1989 the New Technology Telescope arrived, with a 3.58-meter mirror. (We'll have more to say about this one in a moment.)

La Silla (fig. 4.5) was chosen in part because of its extremely dry climate, and it has proved to be an excellent site for infrared observations. In March 2001, for example, the 3.66-meter was used in combination with a new instrument to map the dust that surrounds the innermost young stars in the Orion Nebula, at wavelengths between 5 and 24 microns. As we shall see, the ESO established another major observatory in the Southern Hemisphere, equipped with even larger telescopes.

AN OBSERVATORY DOWN UNDER

For reasons of its own, Great Britain dropped out of the planning for the European Southern Observatory in 1960. The Royal Society of London then approached the Australian Academy of Sciences with a proposal to build and share a large telescope somewhere in Australia. That would give the British the

Fig. 4.5. La Silla Observatory. The large dome at the top houses the 3.66-meter telescope, and the silver box to its front contains the New Technology Telescope.

access to the southern skies that they craved, with the advantage of dealing only with a member of the Commonwealth. The Australians would reap the benefit of a powerful partnership.

In 1967, representatives from the two nations set as their goal a 4-meter telescope, similar to the one Kitt Peak was designing. A formal agreement between the two governments wasn't concluded until 1971. By this time a suitable site had been chosen at Siding Springs, near the small town of Connabarabran in northwestern New South Wales. It lies in rolling countryside at an altitude of 1,100 meters, near a national park with a marvelous Aborigine name, Warrumbungle.

The telescope optics follow the Ritchey-Chrétien design, with a hyperbolic primary and secondary. The primary mirror was cast in Cervit (a low-expansion ceramic) by the now-defunct firm of Grubb-Parsons. It has a usable diameter of 3.89 meters, is 63 cm thick, and weighs a hefty 16 tons. At prime focus the focal ratio is f/3.3, and at Cassegrain focus the ratio is f/7.9.

The telescope was one of the last four telescopes to have an equatorial mount. Its horseshoe equatorial mount is similar to that of the Mayall 4-meter (see fig. 4.2).

Construction of the Anglo-Australian Telescope (AAT) proceeded rapidly once the design was completed, largely because the work was subdivided among many firms in Japan, Switzerland, the United States, Australia, and Britain. Figuring the mirror took four years, from 1969 to 1973. The telescope was inaugurated in 1974 and began regular observations in mid-1975.

In 1998 an Australian-U.K. team began a massive survey of nearby galaxies at the AAT. Named the Two Degree Field Galaxy Redshift Survey (2DF GRS), the project is basically a spectroscopic survey, intended to determine the positions in space of over 250,000 galaxies and to derive such cosmological constants as the mean density of the universe. A map of the distribution of the galaxies shows clearly the large-scale structure of voids and sheets that earlier investigators had discovered.

CARNEGIE MOVES SOUTH

The Carnegie Institution of Washington, established by crusty steel magnate Andrew, has evolved into one of the major sources of funding for science, in particular for astronomy. Recall that the institution had given George Ellery Hale the money he needed to establish the Mount Wilson Observatory. Later, the institution joined with Caltech to operate the 200-inch at Palomar. The Mount Wilson and Palomar Observatories became perhaps the most productive private centers for astronomy in the United States, with some of the best-known scientists on their staff.

As early as 1960 the Carnegie astronomers began to think about a site in the Southern Hemisphere. Like Jurgen Stock, they were attracted to the foothills of the Chilean Andes, which overlook the Atacama Desert. After an intensive search, they settled on Cerro Las Campanas, a 2,400-meter mountain about 100 km north of the city of La Serena. A 1-meter Ritchey-Chrétien telescope, named for Irénée Du Pont, began operating in 1971. It was followed in 1977 by the Henrietta Swope Telescope, a Ritchey-Chrétien 2.5-meter. Swope, one of the few female astronomers at the Carnegie Institution, was famous for her discoveries of variable stars and her studies of their characteristics. The telescope has a classical fork mounting. Its primary mirror is made of fused quartz and has a rather slow focal ratio of f/7 but a wide field of view (1.95 degrees). The Swope Telescope remained the largest on the mountain until the Magellan Project, which we will describe in a later chapter.

AN OBSERVATORY ON A VOLCANO

In the early 1960s, NASA's Apollo program to land a man on the Moon was in full swing. To find a suitable landing site, NASA launched a series of Ranger probes to the Moon. These robots sent back beautiful images, a thousand times more detailed than anything Earth-bound astronomers had ever made. Gerard Kuiper, a Dutch-born American astronomer, led a team that was analyzing these images and advising NASA on places to land. He had quit his position as director of the Yerkes Observatory to establish the Laboratory for Lunar and Planetary Research at the University of Arizona, in Tucson.

Kuiper was blunt, energetic, and supremely self-confident. He was a demanding taskmaster who drove his team to work as hard as he did. He had a considerable reputation as a planetary astronomer, discovering, for example, the atmosphere of Titan, Saturn's giant satellite, and Uranus's satellite Miranda. In 1951 he predicted that a belt of minor planets existed beyond the orbit of Neptune. (It was finally discovered in 1992 and named after him.) His studies of the lunar images proved that the craters had been caused by meteor impacts, not by volcanism, as most others thought.

Kuiper was also a pioneer in the use of infrared detectors to observe the planets. In the early 1960s he began to think about setting up a telescope at a very dry site for infrared studies. Where could he find one? First he prospected in Chile, and it was he who drew Jurgen Stock's attention to the aridity of the Chilean foothills. But before making a decision, he decided to investigate the Hawaiian Islands. He made some tests with a 30-cm telescope at the summit of Haleakala, a 3,000-meter extinct volcano on the idyllic island of Maui. He found some very good seeing there, and the night sky was beautiful. But Haleakala wasn't quite high enough above the normal inversion layer (note 4.7) to escape being occasionally covered by clouds. Off in the distance Kuiper could see much taller volcanoes.

The "Big Island" of Hawaii lies to the southeast of Maui and was formed by five shield volcanoes, of which two (Mauna Loa and Mauna Kea) are over 4,200 meters high. Mauna Kea is the highest island mountain in the world, rising

Fig. 4.6. Several large cinder cones near the summit of Mauna Kea. The largest cone rises 100 meters above the volcanic plateau.

9,750 meters from the ocean floor. In winter the summit of this dormant volcano is covered with snow. (Mauna Kea means "white mountain" in Hawaiian.) Nevertheless, the air at the top is extremely dry.

Kuiper was intrigued by the possibility of an observatory at these heights. He had himself driven to the summit of Mauna Kea, where he found a broad expanse of bare gray lava. Near the top lies a series of giant cinder cones up to 100 meters high. The alpine climate near the summit ensures that the cinders are cemented together by permanent ice. So a cone could provide a stable platform for an observatory. Figure 4.6 shows several typical cones.

After a few tests Kuiper was tremendously enthusiastic. The low humidity on the mountain would guarantee a good infrared signal, and the seeing was outstanding. Kuiper decided to bring a telescope to the mountain. He persuaded the governor of Hawaii, James Burns, to have a jeep trail bulldozed to the summit of Mauna Kea. Then he had a modest observatory with a dome built on Puu Poliahu, one of the cinder cones. He set up his 30-cm telescope and began to make observations of the planets.

After a few months Kuiper was convinced that Mauna Kea would make a perfect site for a large telescope. True, it lacked all the infrastructure, the road, power, and housing that observers would need. And in the thin air at over 4,200 meters his observers found it difficult to breathe, let alone think. Nevertheless, he reported the advantages of Mauna Kea to his contacts at NASA and proposed that NASA should fund a large telescope for his Lunar and Planetary Laboratory.

NASA was receptive to the idea because in the early 1960s the agency was deep into planning a major exploration of the solar system and needed the best information on the planets that it could assemble. But administrators at NASA decided to open the opportunity to competitive bids. John Jefferies, a professor of physics at the University of Hawaii, learned about the plan and decided to submit a proposal on behalf of the university.

Jefferies had only recently arrived in Hawaii. He had made his reputation in solar coronal physics, analyzing observations obtained with a colleague, Frank Orrall, at Sacramento Peak Observatory, and planned to set up a program of solar research on Haleakala. He had no experience in either planetary or stellar astronomy, nor did any of his colleagues at the university. But he could recognize an opportunity when it knocked on his door. He realized that he would have no trouble attracting qualified nighttime astronomers to the university if a first-class telescope were in place on Mauna Kea. So he submitted a proposal for a 2-meter telescope. It would serve the needs of NASA as well as providing an instrument for research and instruction at the university.

Large telescopes are not normally awarded to an institution lacking seasoned astronomers. But somehow in this case the university's proposal was accepted, to Jefferies's surprise and Kuiper's fury. Kuiper "felt [that] 'his mountain' was 'stolen' from him" (quoted from Professor Walter Steiger's web-based memoir on the history of astronomy in Hawaii, www.ifa.hawaii.edu/users/steiger/epilog.htm). In the end, he decided to abandon his Mauna Kea site and concentrate instead on a 1.5-meter telescope on Mount Bigelow, in easy commuting distance from Tucson. It too was funded by NASA.

Jefferies's first task was to find the best location on the mountain for the telescope. The available area is vast. Below the summit, at altitudes between 3,000 to 4,000 meters, a lava plateau extends over 65 sq km. It had been covered with a giant glacier during the last Ice Age, and now its surface is covered with glacial moraines of rough lava boulders. At least half a dozen cinder cones would have to be examined as well, given Kuiper's experience on Puu Poliahu.

Jefferies hired a crew of hardy college students and put them in the charge of Jim Harwood, a lean blond chap who had a mixed background in physics and geology. A rough camp was set up at Hale Pohaku, at an elevation of 2,800 meters, where the observers could acclimatize to the altitude and sleep during the day. A few small trees, low bushes, and grass soften the landscape there.

Each night a group of four to six observers drove a four-wheel-drive truck over the bulldozed track, up the mountain. They were equipped with 10-cm Questar telescopes, which were rugged and portable. They would make simultaneous visual observations from a pair of sites of the same star and record their estimates of the seeing on a scale of 1 to 5. These early observations were sufficient to eliminate the poorer sites. The best sites did indeed seem to exist on top of the cinder cones, especially near the summit.

In time, a double-beam telescope was borrowed from Kitt Peak, and a copy was hurriedly built and put into service. These two photographic instruments provided objective, quantitative measurements of the seeing that a meter-class telescope would experience. The film was sent back to Honolulu to be developed, and the star trails were measured there as well.

In addition to the seeing observations, the observers measured wind speed, temperature, humidity, and cloud cover. Fixed weather stations were set up at a few key points on the mountain. Several small hand-held photometers were used in sunlight to measure the transmission of the atmosphere in the near infrared. (Water vapor blocks much of this radiation.) In addition, an advanced infrared photometer, borrowed from James Westphal at Caltech, was used at night.

The survey involved a lot of hard work and a certain amount of danger. In winter the observers had to wear heavy padded outer clothing and thick "Moon boots," which made clambering over the rocks difficult. Many an ankle was twisted in the course of the two-year survey, but nobody was seriously hurt. On a Moonless night it was easy to get lost, so the observers worked in pairs and used walkie-talkies to keep in touch. The team enjoyed the adventure, recognized the importance of their work, and maintained high morale throughout.

A final report to NASA confirmed Kuiper's experience and intuition: Mauna Kea was indeed a superb astronomical site. The combination of thin air, a surrounding sea, and a steady trade wind combined to produce excellent nighttime seeing. The arid climate would allow observations throughout the infrared.

Jefferies's proposal to NASA called for the construction of a 2.24-meter telescope on a ridge near the main summit. The telescope would have Ritchey-Chrétien optics with a focal ratio of f/10 at prime focus and f/33 at the Coudé focus. (The Coudé focal plane is stationary, outside the telescope, and is used for spectroscopic work.)

The State of Hawaii agreed to build an all-weather road to the summit. At first all electrical power would have to be supplied with generators; a permanent power line would have to wait several years. In addition, construction of a permanent dormitory and dining hall began at Hale Pohaku, at the comfortable altitude of 2,800 meters.

Construction of the telescope began in 1967, under the supervision of engineer Hans Boesgaard, a jolly Dane. First light was seen in 1970.

Jefferies was appointed as director of a new Institute for Astronomy at the university and began to recruit a small army of nighttime astronomers. Mauna Kea was launched.

MAUNA KEA TAKES OFF

By the early 1970s the news was out. Mauna Kea was proving to be one of the best astronomical sites in the Northern Hemisphere, comparable to Cerro Tololo and La Silla in the Southern. Astronomers all around the world began to think about placing telescopes on the mountain.

The Centre Nationale de la Researche Scientifique, the French equivalent of the U.S. National Science Foundation, soon joined forces with the National Research Council of Canada. Together they would build a 3.6-meter telescope just 100 meters closer to the summit than the 2.24-meter Hawaiian telescope. The Institute for Astronomy, as the trustee of the mountain, would receive 10 percent of the available observing time on the new telescope. This kind of arrangement became standard on all future telescopes, to the enormous benefit of the University of Hawaii.

In Honolulu, the governor and the legislature were naturally eager to see this kind of development in Hawaii and set aside a large area at the summit as a scientific reserve. But some resistance began to arise in Hilo, the main city.

Native Hawaiians protested that the summit cone of Mauna Kea is a site sacred to their ancestors and should not be despoiled, even for science. Environmentalists were concerned that a variety of rare birds on the mountain could be threatened as well. And ordinary citizens disliked the prospect of stark white domes that would be visible from Hilo.

Jefferies held a series of town meetings to inform and persuade the people of Hilo. In time he convinced them that the economic and scientific benefits, as well as the prestige Hawaii would enjoy, would outweigh the costs. A master plan for the development of the mountain would be drawn up, with limits on the number and placement of new telescopes.

In time the Canada-France telescope project was approved and funded by the two governments. Construction began in 1976, and first light was achieved in 1979. The telescope has the same yoke construction as the Palomar 200-inch (fig. 4.7). It provides focal ratios of f/3.8 at prime focus, f/8.0 at Cassegrain focus, and f/20.0 at the Coudé focus. Originally intended for photography, the telescope has an unusually wide field of 1 degree at prime focus with a corrector plate.

Fig. 4.7. The 3.6-meter Canada-France-Hawaii Telescope on Mauna Kea has a horseshoe equatorial mount.

During the late 1970s three more large telescopes were built on Mauna Kea to take advantage of its excellent characteristics for infrared astronomy. NASA built its own 3-meter Infra Red Telescope Facility (IRTF) in a compact building. UKIRT, the United Kingdom Infra Red Telescope, a 3.8-meter, was completed in 1979. It is the largest telescope in the world devoted solely to infrared observations. Its mirror could have rather relaxed tolerances of smoothness, because it was intended only for long wavelengths. That factor reduced both fabrication time and cost.

The James Clerk Maxwell Telescope is an impressive 15-meter dish, designed for research at wavelengths between 0.35 and 2.0 mm, which is a transparent window in the atmosphere. The United Kingdom, Canada, and the Netherlands created a consortium to build it. Not to be outdone, Caltech built its own 10-meter dish on the mountain for submillimeter research.

So as of 1979, seven telescopes with mirrors 3 meters and larger were operating on Mauna Kea. Much larger ones would arrive later.

THE SIREN SONG OF THE CANARIES

Despite the success of the Canada-France-Hawaii Telescope, many European astronomers were wary about placing their new telescopes on Mauna Kea. Hawaii seemed rather far away. Construction was expensive, and working at 4,000 meters could be arduous. Wasn't there someplace else, closer to Europe, that might be suitable?

In the 1960s no one was sure where to find the best seeing or even what to look for. The sites already known to have good seeing were all quite different. There was Kitt Peak, a mountain in a desert. There were Mount Wilson, La Silla, and Cerro Tololo, mountains facing a cold sea. And there was Mauna Kea, a tall volcano on an island in a warm sea.

Around 1970, astronomers at Lick Observatory were hoping to build a second 3-meter telescope at a darker site. Merle Walker, a Lick astronomer, carried out a thorough comparison of the seeing in Arizona, Baja California, California, Chile, and Australia. The only common factor among all the good sites was a smooth flow of air. Although Walker tested only these continental sites, he commented that high islands in warm seas, such as the Hawaiian and Canary Islands, also had reports of good seeing. When Hermann Bruck, Astronomer Royal for Scotland, proposed building a Northern Hemisphere Observatory for the United Kingdom, Walker's report was influential in drawing attention to the Canary Islands.

The Canaries, owned by Spain, consist of a fleet of high volcanic islands off the west coast of Africa. Pico de Teide is a 3,700-meter volcano on the island of Tenerife. Way back in 1856, Charles Piazzi Smyth had tested the seeing and sky clarity at Teide and reported excellent results. On the island of La Palma, the 2,400-meter Roque de Los Muchachos lies on the rim of a large volcanic caldera. It too was a potential site, but nothing was known about its qualities.

Then in 1972 a group of solar astronomers arrived at the Roque to test the daytime seeing. They had formed the Joint Organization for Solar Observations (JOSO), with members from more than a dozen European countries. They hoped to find the ideal location for a large solar telescope, and for ten years they had scoured the Mediterranean for the best location. Their problem was not so much one of finding a good site but of reaching a consensus and obtaining government support. Among the many islands they tested were the Canaries, in the Atlantic Ocean. In 1972 a JOSO team found excellent seeing at the Roque. That boded well for nighttime observations, when the atmosphere would be more even more stable.

So later that year a team of British astronomers arrived in the Canaries. They compared the nighttime seeing at the Roque and at Izaña, a settlement at 2,400 meters on the slopes of Teide (fig. 4.8). At each place they photographed star trails around the North Celestial Pole (note 4.8), as Walker had done in his surveys.

The Roque was a difficult place to reach. The observers had to walk several miles over rough lava fields, bring in their equipment by mule, and live in tents. The village of Izaña was luxurious by comparison, but the Roque seemed to have the better seeing.

At this point the Spanish learned of the British presence and were annoyed that they hadn't been consulted. The Spanish awoke to the possibility that they had a valuable asset in the Canaries and decided to take control of it. Site testing by the British was terminated abruptly. Two years passed while the political problems were worked out.

In 1974 the Spanish government convened an international conference to discuss a joint program of site testing. Representatives from Germany, Denmark, Sweden, and the United Kingdom attended a meeting at the University of La Laguna. The Spanish said they wanted a fraction of the observing time on every telescope the foreigners built in the Canary Islands. In return, the

Fig. 4.8. The German Vacuum Tower Telescope *(left)* at the Teide Observatory in Tenerife, Canary Islands, Spain

Spanish agreed to provide the support facilities for the Roque, including a road, power, and water. But an agreement had to wait until the site survey results were complete.

The testing of the Roque site was resumed and continued from 1974 to 1975. The results compared quite favorably with those from Mauna Kea. Both the British and the JOSO team decided for the Roque as the more convenient site. Meanwhile, the design of a large telescope and several smaller ones began.

The British found a ready-made 4.2-meter Cervit blank at the Owens-Illinois Company, left over from the construction of the Anglo-Australian Telescope. That set the scale for their major instrument, to be named the William Herschel Telescope. In time, as the cost of the project grew, the British invited their Dutch and Irish colleagues to participate as full partners. They would share time on the telescope according to their financial contribution.

Negotiations between the Spanish and their prospective tenants ground on.

Finally, in 1979, a firm agreement was signed with the Spanish government. The Spanish would allow Denmark, Sweden, and Britain to build at the Roque if the Spanish would receive 20 percent of the time on all telescopes.

To utilize all this valuable telescope time, the Spanish created a new astronomical institute at Tenerife, the Instituto de Astrofísica de Canarias (IAC). The faculty at the University of La Laguna was also expanded, and the Spanish began an intensive program of training astronomers. These young people would have access to excellent facilities at a world-class site.

The Roque de Los Muchachos has become a major astronomical center since 1979 (fig. 4.9). The British moved their 2.5-meter Isaac Newton Telescope from England to La Palma in 1984. Their pride and joy, the 4.2-meter William Herschel Telescope, was completed there in 1989.

Norway, Denmark, and Sweden joined to build the 2.6-meter Nordic Optical Telescope, which began taking observations in 1989. Then in 1997 the Italians completed Galileo, their 3.58-meter National Telescope. And as we shall

Fig. 4.9. Panorama of Roque de Los Muchachos Observatory, La Palma, Canary Islands, Spain. *From left to right:* the 2.5-meter Isaac Newton Telescope, the Dutch Open Telescope (solar), and the Swedish Solar Telescope.

see in chapter 8, the Spanish are building the Gran Telescopio Canarias, a 10.4-meter telescope at the Roque.

Solar astronomers are also active at the Roque. The Swedes built the Swedish Solar Telescope, an evacuated telescope with a 0.6-meter mirror, and the Dutch built an open solar tower, the Dutch Open Telescope (DOT). German solar astronomers opted instead for Izaña, where they built a Gregorian telescope with a 1.5-meter mirror and a Vacuum Tower Telescope (0.7 meter). The French and Italians joined them there, with a 0.9-meter telescope (THEMIS), especially designed to measure solar magnetic fields. Beginning in 1972, British astronomers from Imperial College observed the infrared sky from the Roque with a 1.52-meter telescope. In 1979 the telescope was donated to the Spanish, who later renamed it the Carlos Sánchez Telescope.

In short, the Canaries now contain two major centers for daytime and nighttime astronomy.

ADVANCED-TECHNOLOGY TELESCOPES

While these new observatories were being established, a quiet revolution in telescope technology was taking place. We can appreciate the new wave by comparing two telescopes that were completed in the same year, 1979. They were 5,000 km apart in distance and a generation apart in concept. On Mauna Kea, the Canada-France-Hawaii-Telescope (CFHT) was the latest example of a proven design that stretched back to the Palomar 200-inch. On Mount Hopkins, 55 km south of Tucson, Arizona, the Multi-Mirror Telescope (MMT) was a collection of radical and risky innovations.

The CFHT had an equatorial mount, similar to the Palomar giant (see fig. 4.7), a thick mirror 3.6 meters in diameter, and a 30-meter dome. Although the CFHT was conventional in these respects, it had several novel features. Two interchangeable secondary mirrors and two Cassegrain foci were provided to optimize observations in the infrared. A three-element corrector plate at prime focus could be used to obtain a full 1-degree field of view. Also, cooling pipes in the floor of the dome were installed to prevent the air from heating up during the day and spoiling the seeing at night.

The telescope performed beautifully even in its first trials. It produced images with sub-arcsecond resolution in the best conditions at Mauna Kea. The sponsors were delighted, but in many ways the CFHT was the last of its breed.

Fig 4.10. The original Multi-Mirror Telescope combined the light from six mirrors, to give it the collecting area of a 4.4-meter telescope.

The Multi-Mirror Telescope was a joint venture of the Smithsonian Astrophysical Observatory, based in Cambridge, Massachusetts, and the University of Arizona. They built the MMT as a test bed for some new ideas and as an attempt to acquire a 4-meter telescope cheaply. The MMT took its name from six circular mirrors, each 1.8 meters in diameter, that were arranged in a hexagonal pattern (fig. 4.10). The light from each telescope was reflected to a common focus on the central axis. That configuration gave the MMT the collecting area of a single 4.4-meter mirror and made it effectively the third largest in the world.

The mirrors had been fabricated with "egg crate" cores to make them light, rigid, and responsive to temperature changes. To keep them aligned and sharing a common focus, they were supported by a computer-controlled system of pistons. Ultimately, the designers wanted to be able to obtain the full resolution of a single 4.4-meter mirror from their set of six smaller mirrors. To do that they planned to use the six mirrors as an optical interferometer, a development we'll describe in chapter 10.

The MMT was one of the first optical telescopes to use a computer-controlled alt-azimuth mount instead of the usual equatorial. In this type of mount the axes are vertical (to turn the telescope in azimuth) and horizontal (to turn in altitude above the horizon). Such a telescope typically has two *Nasmyth foci* (note 4.9) on the platform that turns about the azimuth axis, in addition to the usual prime and Cassegrain foci. Alt-azimuth mounts are more compact than equatorials and therefore allow the use of smaller domes.

This telescope had no dome, however. Instead it was housed in a boxy enclosure that was designed to open wide to the nighttime air. The enclosure contained the computers, the control room, an office, and even a bathroom, all of which rotated with the telescope. This whole package could be much smaller than a conventional dome because of the short focal lengths of the mirrors and the compact alt-azimuth mount. Warm air in the enclosure was sucked out with fans and exhausted far from the telescope, in order to preserve the good seeing.

All these innovations delivered a powerful instrument at a greatly reduced cost, just as the sponsors had hoped. On the other hand, the MMT required a fair bit of fine-tuning before it could demonstrate its full capabilities. Jacques Beckers, the director of the observatory and a highly innovative scientist as well, put in six years of hard work to transform the MMT into an optical interferometer. In 1986 Beckers and his colleagues Keith Hege and J. Hebden succeeded in phasing all six mirrors, so that they performed as a single large mirror. In the end, the MMT became the model for other so-called advanced-technology telescopes, and eventually for the huge telescopes built in the 1990s. As we shall see, the idea of combining several small telescopes to obtain a large collecting area eventually lost favor as mirrors larger than 4 meters became available. The idea survived, however, in the quest for high angular resolution, as chapter 10 describes.

The new technology incorporated into the MMT was not invented solely by the MMT team. Several institutions were developing similar designs, and everyone was exchanging ideas at conferences and in the journals. But the MMT was the first demonstration of a complete system.

High-tech telescopes, similar in most respects to the MMT, were built at several observatories in the 1980s. The Australians at Siding Springs Observatory built the Advanced Technology Telescope early in the decade. It had a single 2.3-meter mirror, an alt-azimuth mount, and a rotating building. In 1987 the 4.2-meter William Herschel Telescope was dedicated at the Roque de Los Muchachos Observatory. And in 1989 the European Southern Observatory completed the New Technology Telescope at La Silla. This telescope was the first to employ a computer-controlled system of actuators (active optics) to maintain the shape of the thin 3.6-meter mirror. The novel enclosure has a sys-

Fig. 4.11. The William Herschel Telescope. Note the two Nasmyth foci on the platform on either side of the telescope, where large heavy instruments can be mounted.

tem of flaps to control the airflow around the mirror. All sources of heat near the telescope, including the motors and electronics, are water cooled to maintain the ambient temperature near the telescope.

The William Herschel is a fine example of an advanced-technology telescope (fig. 4.11). It has four focal planes, an f/2.8 primary focus, an f/11 Cassegrain, and two f/11 Nasmyth ports. Its main instruments include medium- and high-resolution spectrographs, optical and infrared imaging systems, and a wide-field multiobject optical fiber spectrograph. The alt-azimuth mount causes the images at all the foci to rotate as the telescope tracks an object. To eliminate this rotation during a long exposure, instruments at the Cassegrain and prime foci are mounted on counterrotating turntables. This trick doesn't work at the Nasmyth foci, so special optics called *derotators* are inserted in the light beam.

Four-meter telescopes like the Herschel compete effectively with the latest 6- and 8-meter Goliaths—even in the field of cosmology, where light-gathering power is at a premium—by employing state-of-the-art focal plane instruments. At the Herschel Telescope, for example, astronomers from the European Southern Observatory are using S-CAM 2, a *superconducting tunnel junction camera* (note 4.10) to determine the redshifts of quasars as far as 6 billion light-years away. For the first time in the history of astronomy, this camera determines both the time of arrival and the energy of *individual* visible photons. This is a far more efficient way of recording a spectrum than the conventional method with a spectrograph.

THE HUBBLE SPACE TELESCOPE

Throughout the years when the great new observatories were being built around the world, a telescope of a different kind was slowly coming together. Small by contemporary standards, it would outperform all other telescopes for at least a decade.

The Hubble Space Telescope (HST) was launched on April 24, 1990. For fifteen years it has relayed a series of images of the universe that have captivated the public, young and old alike. More important, this instrument has helped astronomers make some astonishing advances in subjects as diverse as cosmology and the search for livable planets. The reasons that the HST has been so productive are not hard to find. It has two great advantages over its brethren here on Earth. At an altitude of 600 km, the HST sees ultraviolet and infrared light undiminished by the Earth's atmosphere. Second, because there is no atmospheric turbulence in orbit, the HST's images are as nearly perfect as its optics allow.

Every great enterprise requires a champion for its success. For the 200-inch (5-meter) telescope, that champion was George Ellery Hale. For the HST, it was Lyman Spitzer Jr. Spitzer is probably less well known to the public than the flamboyant Hale but was every bit as well respected by the astronomical community. He had a dream, fought for it, saw it realized, and was able in the end to enjoy some of its fruits. But he had to wait for over forty years to see it happen.

Spitzer studied physics at Yale University and graduated with a bachelor's degree in 1935. After a year at Cambridge University, he entered Princeton University to study astrophysics under the great Henry Norris Russell. He received his doctorate in 1938, did some teaching, and during World War II worked to develop sonar. After another stint of teaching and research, he returned to Princeton in 1947, where he assumed the mantle of Russell as chair of the astrophysics department. He remained at Princeton for the rest of his life as chair, director of the observatory, and professor.

Spitzer was interested in many topics in astrophysics, but his most important work was in the field of the interstellar medium and star formation. For most of his career he studied how interstellar gases are heated and cooled, with particular attention to the roles of dust and interstellar magnetic fields. He investigated how star clusters might form and how their members would interact. His astronomical research led him into a new field, plasma physics, to which he contributed some fundamental insights. He published a little book on the microphysics of plasmas, *The Physics of Ionized Gases,* that is still considered a classic. He helped to establish the Princeton Plasma Laboratory, whose goal in the 1960s and 1970s was to build a thermonuclear fusion reactor for electrical power (he was the laboratory's first director).

Although primarily a theorist, Spitzer was acutely aware of the importance of critical observations. When he learned of Richard Tousey's pioneering observations of the Sun's ultraviolet spectrum using V2 rockets, he immediately grasped the potential of astronomical observations from space. He began to think about a full-blown observatory in space. A meter-class telescope in orbit would be able to see the full spectrum, from the ultraviolet to the infrared, and, free from atmospheric turbulence, could produce sharper images than ever seen from Earth. In this vision, he was far ahead of his time.

In 1946 Spitzer wrote a report for Project RAND, a research group funded by the U.S. government, in which he outlined the scientific programs a large reflecting telescope might undertake from space. The projects he described might have been written yesterday. They included measuring the distances of faint galaxies, analyzing the structure of galaxies and of globular clusters, and studying the surface and atmosphere of planets. However, his report was probably filed and forgotten until 1957, when the Russians launched Sputnik and the U.S. government was shocked into action. This was the height of the cold war, in which a nation's technical ability was considered an indicator of its success as a society.

NASA, the National Aeronautics and Space Administration, was hastily organized by Congress in 1958 and immediately laid plans to observe the brightest object in the sky, the Sun. Between 1962 and 1975 a series of eight Orbiting Solar Observatories were launched, with tremendous benefits for solar and high-energy astrophysics. Skylab, a well-instrumented solar observatory, made major discoveries about the solar corona during a nine-month flight in 1973.

At the same time, NASA did not neglect the interests of nighttime astronomers. Four Orbiting Astronomical Observatories were launched between 1966 and 1978. The first one died of a tired battery three days after reaching orbit, and the fourth never reached orbit. But OAO 2, launched in 1968 and carrying a cluster of eleven ultraviolet telescopes, worked beautifully for five years. It saw a supernova and the center of Messier 31 and enabled astronomers at the Smithsonian Institution and the University of Wisconsin to explore the ultraviolet spectra of hot stars.

Spitzer decided to propose an experiment for one of the Orbiting Astronomical Observatories. In the late 1960s he pulled together a team that built an instrument for OAO 3, later named Copernicus. It included a 0.8-meter telescope, the largest deployed in space up to that time, and a spectrograph that covered the important ultraviolet spectrum of hydrogen between 90 and 120 nm. (This spectrum is appropriately named the Lyman series—not for Lyman Spitzer but for Theodore Lyman, who discovered the series in 1908.) An x-ray telescope built by the University College of London rode alongside the Princeton telescope. Copernicus was launched in 1972 and continued to pump out data until 1981. Spitzer's team used it to study the UV spectrum of interstellar gas and dust, as well as the atmospheres of hot stars.

The OAO program was, on balance, a great success and helped to convince conservative optical astronomers that they had a future in space. But progress was slow. Spitzer continued to push for a large telescope in space all through the 1960s and 1970s.

In 1965 NASA asked the National Academy of Sciences to recommend future space programs. The academy set up a blue-ribbon study group, which Spitzer was asked to join. At a crucial meeting at Woods Hole, Massachusetts, he persuaded the group to endorse a meter-class, general-purpose telescope. It should have the resolution of a ground-based telescope of 3 meters (say, 0.05 arcsecond at 500 nm) and be capable of detecting wavelengths between 80 nm and 1 mm (not an easy job, even today). The group urged NASA to start designing the telescope in 1968 and to launch it in 1979.

Spitzer then joined the National Academy's Space Science Board, which met over the next seven years with many consultants. It examined the engineering problems that might arise in building the Large Space Telescope, as it was called, and saw nothing insurmountable. It also recognized that the telescope, built to last at least a decade, would require occasional maintenance by astronauts, as well as updating of its instrument packages. The board also laid out a detailed scientific program for the telescope, which was intended to convince more astronomers that the project was worth pursuing.

During this period, the Large Space Telescope (LST) was studied and endorsed by a growing number of elite committees. Many astronomers still worried, however, that such a large project was premature. They thought that more science could be achieved with less risk and lower costs with smaller instrument packages. So in 1978 NASA launched the International Ultraviolet Explorer (IUE), a joint effort with the European Space Agency (ESA) and the Science Committee of the United Kingdom. This satellite carried a 45-cm telescope, two spectrographs covering the range from 115 to 335 nm, and a cluster of cameras. Launched in 1978, IUE was the first orbiting observatory to offer "visiting" scientists opportunities to observe from space on short notice. IUE was enormously successful, operating for over eighteen years despite losing its six gyroscopes one by one. Among its discoveries were the auroras of Jupiter, high-speed winds from young stars, and "star spots" analogous to sunspots.

Meanwhile, NASA engineers continued to design the details of the Large Space Telescope and its instruments. As the design progressed, the projected cost of the telescope rose to $500 million. At that point, Congress became alarmed. In 1974 the House Appropriations Committee deleted a $6 million appropriation for continuing the design. Spitzer hastily mobilized a campaign among astronomers to lobby Congress to restore the funds. A storm of letters and calls, along with personal appeals from eminent scientists, persuaded key members of the Senate to restore half of the funds.

Congress was still reluctant to approve the whole scheme. It was not only the cost; the scientific value of the project also eluded the members. As Spitzer wrote, "One Congressman remarked that he had cooperated with the astronomers in obtaining approval for the Very Large Array, which was supposed to unravel many of the riddles of the Universe as a whole, but he was puzzled as to why astronomers were now requesting funds for another expensive instrument with apparently the same purpose" (*Quarterly Journal of the Royal Astronomical Society* 20, 29, 1979).

So Spitzer geared up once again to convince key congressmen. A parade of distinguished scientists testified before the committees, to educate and intrigue the members and to pry loose approval for the LST. Such articulate speakers as John Bahcall (of the Institute for Advanced Study at Princeton) and George Field (professor of astrophysics at Harvard University) were enlisted in the campaign.

At the same time, the project was scaled back to reduce costs and to guarantee success in the building of the telescope. The mirror was reduced from 3 to 2.4 meters, and a more compact configuration of the auxiliary instruments was designed. In addition, some European partners were brought in. The European Space Agency agreed to build the solar arrays and one of the focal plane instruments in exchange for access to the telescope.

Finally, in the summer of 1977, Congress approved the project with an initial appropriation of $36 million. How this infant would grow!

NASA assigned responsibility for the design, development, and construction of the telescope to the Marshall Space Flight Center in Huntsville, Alabama. NASA's Goddard Space Flight Center would oversee construction of the auxiliary instruments and set up the ground control center at Greenbelt, Maryland. From the outset, the LST was designed for launch and servicing with the space shuttle. Manned space flight, with exciting extravehicular activity, was definitely a part of the whole scheme.

Everyone connected with the project recognized the awesome challenge it presented. Every aspect of the telescope would push the envelope of technology. The optics would have to be nearly perfect. The pointing accuracy would be unprecedented, certainly in space. Satellite control, data handling, electronics, and solid-state detectors—all these would have to be improved or invented to fit the needs of this unique observatory.

A MIRROR WITHOUT PEERS

The optical design called for a Ritchey-Chrétien Cassegrain (see note 4.2) with aspheric primary and secondary mirrors to achieve a wide field (3.4 arcmin-

Fig. 5.1. The honey-comb structure of the two mirrors cast in Ultra Low Expansion silica glass for the Hubble Space Telescope

utes) free of coma aberrations. With an effective focal length of 57.6 meters and a mirror diameter of 2.4 meters, the telescope's focal ratio would be f/24. The mirror would be coated to reflect wavelengths as short as 120 nm and as long as 1,200 nm. (Visible wavelengths are around 500 nm.) At ultraviolet wavelengths, the telescope would be able to resolve details as small as 0.02 arcsecond, ten times smaller than any ground-based telescope. Such resolution would allow the LST to detect objects fifty times fainter than any other (note 5.1).

The primary mirror was the key to the whole project. If the telescope were to achieve *diffraction-limited resolution* (note 5.2) at ultraviolet wavelengths, the mirror would have to depart from a perfect paraboloid by no more than one-twentieth of a wavelength and have no ripples larger than one-sixtieth of a wavelength, or 2 nm.

NASA decided to have two mirrors fabricated, a flight model and a backup. The Perkin-Elmer Corporation (now part of Hughes Danbury Optical Systems) was chosen for the first and the Eastman Kodak Company for the second. Both firms had excellent reputations as manufacturers of precision optical instruments. Perkin-Elmer in addition had made superb mirrors for military reconnaissance satellites. The two firms used similar procedures in fabricating the mirrors, except that Perkin-Elmer built a state-of-the-art computer-driven polishing facility, while Eastman Kodak relied on more traditional manual operations by skilled opticians.

NASA selected Corning Glass Works to fabricate two blanks of Ultra Low Expansion silica glass for the mirrors. Instead of casting a solid blank, which would have been too heavy, Corning built a rigid lightweight sandwich from separate parts. A thin faceplate (25 mm thick), a back plate, rings, and an "egg crate" core were fused into a single unit (fig. 5.1). The whole assembly was supported at its center, and the outer rim was allowed to slump during the fusing process. In this way a convex backside and concave front side were created to reduce the amount of grinding that would be needed. Despite this innovative construction, the whole assembly weighed 900 kg, about the same as a sport utility vehicle. Corning delivered the blanks in November 1978. Grinding and polishing would take three years.

Perkin-Elmer devised a clever scheme to support the blank during the grinding operations. Despite its intrinsic rigidity, the blank would droop if it were supported only at its rim. So in collaboration with NASA, the company built an array of actuators that supported the blank at 130 points. The actuators applied exactly the right amount of force to reproduce the shape the mirror would have in gravity-free space.

The finished mirror was 2.4 meters in diameter and 30 cm thick, and it weighed 773 kg, or about one-fifth of what a conventional solid mirror would weigh. The surface was so smooth that if it were expanded to an area the size of Colorado, nothing taller than 0.6 mm would stand up. To preserve the precise shape of the mirror once it arrived in orbit, it was mounted in a special cell equipped with an array of actuators. These could be adjusted in flight to compensate for any deformation of the mirror.

Before being coated, the surface of the mirror had to be thoroughly cleaned. Even a single submicron dust particle adhering to the surface could scatter enough light to contaminate an image. NASA enlisted a company whose specialty was cleaning microcircuits, and their method was adapted to cleaning the large mirrors (see note 5.3 on dirt in space).

At Perkin-Elmer a huge vacuum chamber was built to evaporate coatings onto the mirror. A layer of aluminum 8 microns thick was first applied. Then an overcoat of magnesium fluoride 2 microns thick was added to prevent oxidation of the aluminum and to improve the reflectivity at ultraviolet wavelengths.

POINTING THE TELESCOPE

NASA engineers couldn't plan to point the spacecraft using gas thrusters, because the exhaust gas would linger nearby and spoil observations. So the craft was provided with a set of gyroscopes, three flywheels (one for each direction, pitch, roll, and yaw), and a set of three fine-guidance sensors in the focal plane. To point the telescope, controllers command the flywheels to spin at a specified speed for a specified time. The 12-ton spacecraft moves (slowly!) in reaction to the forces generated by the flywheels. After the telescope arrives at its desired orientation, the fine sensors lock on to two stars in the field of view. Thereafter the gyros sense coarse drifting motions, and the fine-guiders sense the smaller ones.

The sensors are designed to detect a drift of the telescope as small as 0.0028 arcsecond. A sensitive servomechanism, linked to the sensors, cancels the drift and holds the telescope steady, with a "jitter" of only 0.007 arcsecond over long periods. That's like holding a laser beam on a dime at distance of 300 km. And all this goes on as the telescope orbits the Earth at 7 km/s.

For this scheme to work, the coordinates of fairly bright stars in every possible field of view must be accessible to the fine-guidance sensors. So before launch the coordinates of nineteen million stars with magnitudes between 7 and 16 (note 5.4) were measured on the Palomar 48-inch Schmidt Sky Survey and the United Kingdom's Schmidt SERC-J plates. Once a target is chosen, the control center relays coordinates of stars in the field of view to the onboard computer. When the sensors lock on to these reference stars, the telescope is offset a few seconds of arc to the exact location of the target.

THE REST OF THE HARDWARE

The telescope optics and pointing sensors were perhaps the most critical items, but there was a lot of additional equipment required to furnish an observatory. Lockheed Martin Missiles and Space Company was responsible for building the spacecraft that contains, protects, and powers the telescope and allows

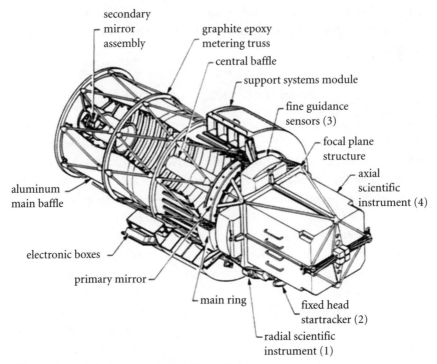

Fig. 5.2. The main features of the Hubble Space Telescope

it to communicate with the control center. You can see some of these elements in figure 5.2.

The telescope optical assembly is mounted in a rigid cylindrical cage made of graphite epoxy resin. This lightweight, strong material was developed just in time for the project. At the rear, behind the focal plane, an array of four instruments is located. (We'll describe these in a moment.) The light beam from the focal plane can be switched to any of these. The solar panels provided by the European Space Agency supply 2.4 kW of power to the electronics, which are housed in crates surrounding the light path. Batteries provide backup power when the telescope scoots through the Earth's shadow. Blankets of heat shields screen all these components from the Sun, and the front end has a door that can be closed to shut out sunlight. The whole rig is contained in a cylindrical tube, and it weighed 12.5 tons at launch.

THE DATA PIPELINE

All the images and spectra obtained by the telescope are converted to digital signals by *charge-coupled devices,* or CCDs (note 5.5). In the 1970s and 1980s these solid-state detectors were developing rapidly in size and wavelength sensitivity. The needs of the LST pushed the pace of development. When they first arrived on the market, CCDs were arrays of at most 128 × 128 pixels. When the telescope was launched, in 1990, it carried CCDs of 1,024 × 1,024 pixels, a sixty-four-fold increase in area.

The Hubble Space Telescope (or HST, the name given to the observatory at launch) was designed to observe for many hours a day, collecting gigabytes of digital data. (Currently the HST produces 3 to 5 gigabytes a day.) How could one get this flood of data down to Earth?

In 1983 NASA had launched the first of three Tracking and Data Relay Satellites into a geosynchronous orbit (note 5.6) to service its fleet of scientific satellites. NASA planned to have the HST send its data to the TDRS system, which would relay it to the control center at Goddard Space Flight Center. However, the HST would not always have a line of sight to one of the communication satellites. The solution to this problem was to store the data onboard until a chance arose to dump it. In the 1970s the biggest, fastest data recorders used magnetic tape, winding from one reel to another. They were heavy, ornery, and, like all mechanical devices, prone to failure. To ensure uninterrupted operation, NASA insisted on installing three of these onboard.

Astronomers from all over the world have access to the HST, but somebody had to decide which observations to make and when to schedule them. So in 1983 NASA created the Space Telescope Science Institute, which is based at the Johns Hopkins University in Baltimore, Maryland. Riccardo Giacconi, the pioneering x-ray astronomer, was appointed the first director. The institute's staff and its consultants review proposals from astronomers who wish to observe with the HST, help design the observations, schedule time on the telescope, and process the data after they are received from the ground control center at the Goddard Space Flight Center. The control center translates the observing requirements into operations the onboard computer understands, recovers the raw data, and monitors the performance of the spacecraft.

THE LONG DELAY

Construction of the HST took eight years, 1977 to 1985. Perkin-Elmer finished polishing and testing the primary mirror in 1981. It was a jewel, more finely polished than any before. ESA, NASA, and the University of Arizona were responsible for building the instruments. They delivered their gems in 1983, and Perkin-Elmer finished the rest of the optical assembly a year later. By 1985 everything had been installed and tested in the spacecraft. The Hubble was ready to fly, and a launch was expected sometime in 1986.

Then disaster struck. On January 28, 1986, the space shuttle *Challenger* exploded shortly after liftoff, killing all seven astronauts aboard. The nation was stunned. NASA shut down all shuttle operations until the cause of the accident could be determined. President Reagan appointed a committee chaired by William Rogers, former secretary of state, which included some of the brightest minds in the country, to investigate.

Richard Feynman, the unconventional Nobel Prize–winning physicist, eventually identified the specific cause of this horrific accident. The culprit was the rubber O-ring that seals the segments of one of the two solid rocket boosters. *Challenger*'s launch had been delayed for various reasons, day after day, for six days. All during this time the shuttle had been exposed to cold weather. The low temperatures caused the O-rings to shrink, which opened a crack. During the firing of the solid rocket boosters, hot exhaust gases escaped from the crack. The gases heated the external hydrogen fuel tank, which exploded some seventy-seven seconds after launch.

NASA responded to the committee report by revising its safety measures and redesigning the shuttle. Those tasks consumed thirty-two months. With no launch in sight, NASA and Lockheed engineers used the delay to test every component of the HST to assure maximum reliability.

IN ORBIT AT LAST

Finally, on April 24, 1990, the Hubble Space Telescope was launched into an elliptical orbit with an average altitude of 600 km. At that altitude, the HST would have a lifetime of at least twenty years (note 5.7). But after a few months of checking out the systems, astronomers at the Space Telescope Science Institute recognized a serious problem. They had expected to see images with ten times the spatial resolution of ground-based telescopes. The HST's images of galaxies were fuzzy, however, and each star was surrounded by an artificial halo of light. Something was definitely wrong with the telescope's mirror.

In time, the mirror's flaw was identified as *spherical aberration* (note 5.8). The edge of the mirror had been ground too flat by one-fiftieth of the width of a human hair, or 2.2 microns. That may sound small, but in optical work it corresponds to 4 wavelengths, which is a huge error. How could this possibly have happened? Perkin-Elmer had carried out exhaustive tests all during the polishing of the mirror. But as a later investigation revealed, the tests themselves were flawed, and warning signs that something was wrong were ignored.

To test the mirror, Perkin-Elmer had used a *reflective null corrector,* which consists of two mirrors, a lens, and a laser. The corrector contains a critical measuring rod. Somehow, somebody had misplaced a cap on the rod, and that introduced a 1.3-mm spacing error in the corrector, which was then incorporated into the shaping of the mirror. But that was only the beginning of the story. The corrector itself was checked for alignment along the axis of the mirror with a so-called *inverse null* before each testing session. These alignment tests indicated that there were problems with the setting of the corrector, but as long as the team was satisfied that the test equipment was *aligned,* it ignored these other problems. In 1981, just before the delivery of the mirror, another test with a different type of device (a *refractive null*) showed that the mirror had a serious spherical aberration. The test team also discounted these results. Midlevel officials in the company were aware that two different tests had given disturbing results, but they chose not to alert their superiors.

In this way the mirror was made with the smoothest surface of any mirror before it, but with the wrong shape. The telescope's performance was seriously compromised. Only the onboard Wide Field Camera gave useful results, and only on the brightest objects. Ironically, the backup mirror built by Eastman Kodak was later found to be perfect. It now resides at the Smithsonian Aerospace Museum in Washington.

RESCUING THE HUBBLE

NASA could not possibly consider bringing the HST mirror back to Earth for repairs or replacing it in orbit. At the same time, over $1.5 billion (to say nothing of NASA's reputation) had been invested in this observatory, so it couldn't be written off as a failure. A solution was quickly found: astronauts would visit the HST and insert corrective optics into the light beam. A crew of seven were trained in a kind of virtual-reality facility to teach them where the different components of the HST were located and how they looked in place. They had to learn to use a battery of special tools that would be needed in the delicate operation.

In the meantime, a package of optics called the Corrective Space Telescope Axial Replacement (COSTAR) was built by Tinsley Laboratories and Optical Research Associates and assembled by Ball Aerospace. This package would replace one of the existing instruments, the High-Speed Photometer. Mirrors were used instead of lenses, because lenses would absorb the ultraviolet light that the HST was built to see. The mirrors were the size of coins, but unlike coins, they were not rotationally symmetric. The optical tolerances for these coin-sized mirrors were also very demanding: the shapes of the mirrors had to be accurate to within 6 nm, and their surfaces had to be smoothed to 1 nm, or one-ten-millionth of a centimeter. When fully assembled, COSTAR contained five pairs of mirrors to correct the beam directed to each of the five onboard instruments.

On the night of December 2, 1993, the shuttle *Endeavour* roared off its launch pad with a crew of seven astronauts, to capture the HST and to install the COSTAR, new solar panels, additional gyroscopes and a replacement of the Wide Field/Planetary Camera. *Endeavour* chased the HST, caught up with it, and maneuvered to allow a robotic arm (the "Canada arm") to pick up the wounded bird and nestle it in the shuttle's bay. The HST has a mass of about 10,000 kg on Earth, and despite the gravity-free environment in space, it has a lot of inertia. Maneuvering it gently into the bay was something like easing a piano through a window, and therefore required a delicate touch.

Once the HST was safely stowed, the astronauts examined it with remote cameras on the arm. Then they began a series of space walks to repair it. Imagine how they must have felt, with the whole world watching and waiting, and with the future of this unique observatory at stake. Astronaut Jeffrey Hoffman replaced the original Wide Field Camera with the improved model, which contained its own set of corrective optics. Astronauts Kathryn Thornton and Thomas Akers installed the COSTAR. This refrigerator-sized crate weighed 300 kg on Earth and required some brute force to maneuver. Once the critical COSTAR was in place, the other components were installed and checked out

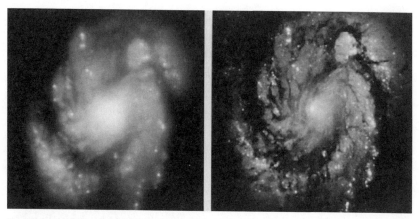

Fig. 5.3. The Whirlpool Galaxy, Messier 61, before *(left)* and after the Wide Field Camera optics were corrected

and the HST was released again. This arduous rescue operation lasted eleven days and required five space walks by two pairs of astronauts. But it worked, thanks to the astronauts' skill and daring. Figure 5.3 shows COSTAR's improvement to the quality of images. This service mission was the first of four and demonstrated NASA's ability to keep the HST in top form and supplied with state-of-the-art instruments.

THE HUBBLE'S EYES

The real working tools of the astronomers, the spectrographs and cameras that analyze and record the light, are clustered around the axis of the telescope, behind the primary mirror (see fig. 5.2). These five instruments are built as modules that can be pulled out easily and replaced with improved versions during the long life of the HST. Each module is the size of a telephone booth, with enough room to contain all the optics and associated hardware. The instruments share the focal plane, which is the size and shape of a dinner plate. COSTAR's optics divide the available light among the five instruments and the three fine-guiders.

The Wide Field and Planetary Camera, as its name implies, is used for imaging broad swaths of the sky, a planet, or a nearby galaxy. WFPC 2, the latest onboard version, was built by NASA's Jet Propulsion Laboratory. It is really a cluster of four cameras that share the light. Each camera is equipped with an

800×800 pixel CCD. Three cameras each have a field of view of 80 arcseconds, while the high-resolution planetary camera has a field of 37 arcseconds. The WFPC 2 has no fewer than twelve filter wheels and forty-eight narrow-band filters that cover the wavelength range from 370 to 1,100 nm. This mechanical wonder allows an observer to choose an optimum bandpass for each camera. The image of the Whirlpool Galaxy in the right-hand panel of figure 5.3 is an example of the high-quality imaging this instrument delivers. A more advanced model, WFPC 3, is under construction.

The Near Infrared and Multi-Object Spectrometer (NICMOS) was built at the University of Arizona and installed in 1997 during the second service mission. This device forms images at wavelengths from 0.8 to 2.5 microns of three adjacent areas, at different magnifications, simultaneously, which allows an astronomer to zoom in on a target. NICMOS uses three mercury-cadmium-telluride infrared arrays, each with 256×256 pixels. These arrays must be cooled to 77 kelvin to reduce the thermal noise from the spacecraft, so in the first version they were implanted in a 100-kg block of frozen nitrogen. Unfortunately, heat leaked into the block, shortening its life from five years to two. Without adequate cooling, NICMOS had to be shut down for more than three years.

NASA initiated a fast-track program to develop an onboard refrigerator (a *cryostat*) to cool the NICMOS with cold neon gas. The cryostat was developed from scratch in fourteen months, a record of sorts, and was installed in 2002 during the third service mission. Incidentally, the cryostat has a turbine that spins at an incredible 400,000 rpm, a hundred times the speed of a car engine. NICMOS recovered nicely and delivered a long list of notable observations of such targets as star-forming regions (fig. 5.4), quasars, and the dusty disks of galaxies.

The European Space Agency had contributed the Faint Object Camera (FOC) to the HST's original arsenal. It was designed to make images of the faintest of galaxies and also to resolve the smallest details on brighter objects. In fact, the camera was the only instrument in the original set able to utilize the full spatial resolution of the telescope, 0.02 arcsecond in ultraviolet light. The device contained an image intensifier that converted a few photons into a heavy shower of electrons, thereby amplifying the light by a factor of one hundred thousand. After COSTAR corrected the FOC's light beam, it was able to detect objects as faint as magnitude 27.5. (That's the brightness our Sun would have at a distance of a million light-years.) To avoid being blinded by the brighter objects in the field of view, the FOC employed a device to block their light.

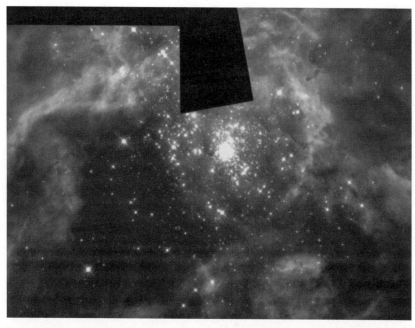

Fig. 5.4. Massive stars in 30 Doradus, a nebula in the Large Magellanic Cloud, generate powerful winds that compress the surrounding interstellar clouds. As a result, new stars are born. Observations from the Near Infrared and Multi-Object Spectrometer at infrared wavelengths pierce the dense dust clouds to reveal this star nursery.

After serving for twelve years, the FOC was replaced in March 2002 by ESA's Advanced Camera for Surveys. Overall, this new instrument is five times as sensitive as the FOC and has twice as wide a field of view. It has three CCD cameras and an assortment of filters and dispersers. The Wide Field Camera is especially sensitive in the near infrared (say, 700 nm) and is useful for searches of faint galaxies. It has a field of view of 202 arcseconds and resolves 0.1 arcsecond. Then there's the High-Resolution Camera, which has a coronagraphic feature to enable a search for a faint object near a very bright one. It resolves 0.05 arcsecond over a field of 29 arcseconds. Finally, a Solar-Blind Camera accepts only the far ultraviolet (say, around 121 nm). With 0.06-arcsecond resolution over a field of 34 arcseconds, this instrument is designed to find hot stars and quasars.

Imaging is important, but astronomers also depend on spectroscopy for much of their science. Two of Hubble's original instruments were the Goddard High-Resolution Spectrograph, intended primarily for use at ultraviolet wavelengths, and the Faint Object Spectrograph. These instruments were replaced in 1997 by the Space Telescope Imaging Spectrograph (STIS), a two-dimensional spectrograph. It focuses five hundred points of an extended object along its linear slit and expands the light from each point into a spectrum. All told, the STIS delivers thirty times more spectral data and five hundred times more spatial data than the Goddard instrument did. STIS captured the spectrum of the expanding shell from supernova 1987A, the brightest supernova seen in the past 350 years (fig. 5.5).

The European Space Agency is developing a radically new photon detector for the Hubble Space Telescope that may in time replace the venerable charge-coupled devices. CCDs have many advantages, including large size (arrays of

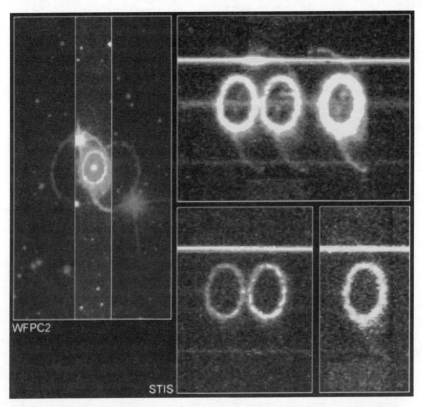

Fig. 5.5. The expanding shell of supernova 1987A was imaged *(right)* in a series of spectral lines by the Space Telescope Imaging Spectrograph. Each ring represents a different element in the gas. For comparison, see the image in visible light *(left)* obtained with the Wide Field Camera.

up to 4,000 × 4,000 pixels) and high sensitivity in the ultraviolet as well as the infrared. Each pixel in a CCD absorbs a few photons and converts them to electrons, which are collected to form a digital image. The conversion from light to electrons is relatively insensitive to the color of the photons. In contrast, ESA's S-CAM, a device based on *Josephson junctions* (note 5.9), detects the position, arrival time, and color of individual photons. In essence, the S-CAM is about a hundred times more sensitive than a CCD, which means that much shorter exposures are needed to reach the same precision in a spectrum. To reduce the thermal noise from its surroundings, the device is immersed in liquid helium at a temperature less than 1 degree kelvin above absolute zero. ESA scientists have tested the S-CAM successfully at the William Herschel Telescope in La Palma. So far the S-CAM arrays are small, only 6 × 6 pixels, but the team is busy building much larger ones.

THE HUBBLE'S TROUBLES

After nine years of outstanding performance, the HST fell silent on November 13, 1999. The fourth of six gyroscopes had failed, and with only two left, the HST couldn't stabilize itself in space. The spacecraft went into a standby (or sleep) mode.

NASA was prepared. After the third gyro failed, the agency hurriedly organized a rescue mission, fitting two flights of the space shuttle into its busy schedule. Service mission 3A was launched on the shuttle *Discovery* in December 1999 with seven astronauts aboard. Working in space is both arduous and dangerous, so four of the astronauts shared "extravehicular sorties" on alternating days. Their first priority was to replace all six gyros. For good measure they replaced a faulty guidance sensor and the main computer. An old mechanical data recorder was also extracted and replaced with a solid-state model. The mission was an unqualified success. The HST woke up and went back to work with renewed vigor. Such service missions are not cheap, however. This one cost taxpayers over $200 million, but it was worth every penny.

Nine years in space had aged the HST. Harsh ultraviolet light, cosmic rays, and the steady rain of small, fast meteoroids had gradually degraded the solar panels that provide power to the spacecraft. So in March 2002 a fourth service mission was sent to the telescope to replace the panels with more robust models. During this twelve-day flight the astronauts also installed ESA's Advanced Camera for Surveys (see page 119).

HIGHLIGHTS

As of 2002, the price tag for the HST, including construction and operation for twelve years, was over $3 billion. Let's see what it had produced.

Few readers of this book can have failed to be impressed by the exquisite images of planets, nebulae, and galaxies that the HST has sent back to us. The Eskimo Nebula (fig. 5.6) is one of my favorites.

Some of the most important scientific results coming out of the HST data do not necessarily produce the prettiest pictures, however. As an example, consider the Hubble Deep Field. One of the big questions in cosmology is how the universe has changed during its lifetime. Were there fewer galaxies in the early universe than now? Did the relative numbers of different types of galaxies change over time? Were there more spirals or irregular galaxies then than now, for example? To find out, one has to look back in time, and that means looking at the most distant galaxies.

So in 1995 the HST stared at a tiny area near the Big Dipper for a full ten days, to obtain images in visible light of the galaxies that lie at the extreme limit of the observable universe. The Wide Field Camera saw over fifteen hundred galaxies in this narrow window (fig. 5.7). In addition to the familiar spiral and elliptical galaxies, there is a zoo of strange types, never seen before. Some of these may have been born only a billion years after the Big Bang.

In 1998 the NICMOS aboard the HST reexamined the same area in the sky. With its enhanced sensitivity to infrared light, NICMOS was able to detect even fainter galaxies, down to magnitude 30. Some of these may be 12 billion light-years away. Astronomers are still analyzing these images and comparing them to a similar set made by the HST of a region in the southern sky.

Another big question facing cosmologists is whether the expansion of the universe is steady, slowing down, or speeding up. Will the universe grind to a halt at some very distant time and begin to collapse? To answer that question, one wants a precise measurement of the present rate of expansion—or, in other words, the present value of the Hubble constant (H_0). The reciprocal of the constant ($1/H_0$) also yields an estimate of the age of the universe, a vital number in cosmology.

As we learned earlier, Edwin Hubble's first estimate of his constant (which is probably not constant at all) was 500 km/s per megaparsec. This value was based on a faulty calibration of the brightness of Cepheid variables. Later work

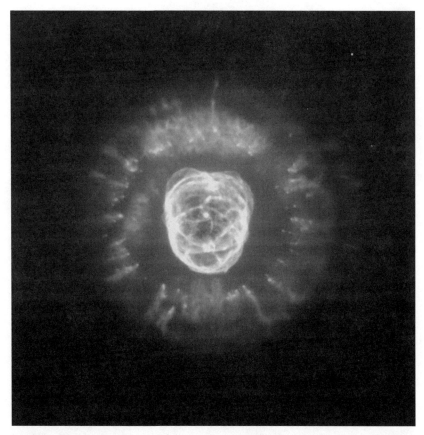

Fig. 5.6. The Eskimo Nebula, NGC 2237, as seen by the Wide Field Camera on the Hubble Space Telescope

has reduced H_0 to something below 100, but with a big uncertainty. So in 1993 Wendy Freedman (Carnegie Observatories) and her international team began the HST Key Project, with the aim of reducing the uncertainty in H_0 to 10 percent or less. What tool could be more appropriate for this purpose than the HST?

Like Hubble, Freedman's group used Cepheid variables as standard candles in relatively near galaxies but checked their distances with five different independent methods to eliminate systematic errors. As of 1996, the Key Project's best value of H_0 was 73, with an uncertainty of about 14 percent. At the same

Fig. 5.7. The Hubble Deep Field, an image of fifteen hundred distant galaxies in a narrow tunnel through space

time, Alan Sandage (also from the Carnegie Observatories) and his colleagues also used the HST to determine H_0. Instead of Cepheids, however, they chose Type Ia supernovas as standard candles. The more slowly these supernovas brighten and dim, the greater is their absolute brightness. Sandage and company obtained a preliminary estimate of about 57 km/s per megaparsec, well below Freedman's value. The corresponding estimates of the age of the universe lie between 12 and 14 billion years. A hot debate has ensued, and, as we shall see, the high and low values eventually converged.

The Hubble Space Telescope has made possible many other important ob-

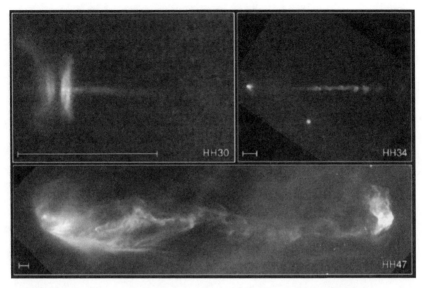

Fig. 5.8. A young star ejects high-speed jets of gas while its accretion disk collapses.

servations, such as the jets and accretion disks that accompany the birth of stars (fig. 5.8), the impact of Comet Shoemaker-Levy onto Jupiter, the velocities of gases near supermassive black holes, and the discovery of many gravitational lenses (fig. 5.9). In a later chapter we'll learn how the HST has teamed up with x-ray and ground-based telescopes to study the far limits of the universe and to catch a gamma-ray burster in full bloom.

THE FATE OF THE HUBBLE SPACE TELESCOPE

In 2004 NASA announced that the HST had far exceeded its planned life span and would no longer be serviced. The decision to let the Hubble die was based in part on concerns for the safety of astronauts, and in part on the president's goal of a manned flight to Mars. However, the public and the scientific community raised such a fuss that NASA requested a panel of the National Academy of Sciences to reconsider the future of the HST. In July 2004 the panel concluded that the HST could still produce much good science and that NASA should consider alternative servicing modes, perhaps with robots.

NASA undertook a study of servicing the HST with a robotic mission. The

Fig. 5.9. The sharp arcs in this picture are multiple images of a single distant galaxy whose light passes through the gravitational lens of the bright galaxies at the center of the photograph.

results were disappointing; the difficulty of maneuvering a robot, with low risk to the telescope, was extremely high. In its final report the National Academy committee agreed that robotic servicing was impractical, but that a robot could be used to cause the Hubble to crash in an ocean, far from populated areas.

President George W. Bush's 2006 budget includes no funds for a servicing mission with the space shuttle. But Michael Griffin, the newly appointed administrator of NASA, declared in April 2005 that he will reassess the feasibility of a shuttle servicing mission once the shuttle fleet is back in operation following the *Columbia* disaster.

So, as of this writing, the future of the Hubble is still uncertain.

A MIRROR IN MANY PIECES

ASTRONOMERS ARE used to dealing with unconventional ideas, even some that seem fantastic. After all, that's what makes astronomy exciting. But when it comes to building telescopes, they are conservative. Each new telescope involves a lot of money, and money is not something you risk casually. So in building telescopes, astronomers tend to follow the same well-tested designs—until, that is, they bump up against a solid wall of cost. Then the door opens to a new idea.

That is essentially what happened around 1970, when California astronomers began to think about a successor to the 200-inch telescope.

A TENTATIVE BEGINNING

Thanks to George Ellery Hale, California led the world in building large telescopes, and in using them to probe the far reaches of the universe. The Palomar Observatory with its 200-inch (see fig. 2.4) was preeminent. By agreement, Caltech owned it and shared it only with astronomers at the Mount Wilson Observatory. As cosmology blossomed into an exciting field of astronomy, astronomers at other institutions found themselves at a disadvantage.

Professional astronomers were not the only ones hungry for big-telescope time. The huge expansion in the scope of astronomy that occurred during the 1960s and 1970s attracted large numbers of students to the field. As we saw in chapter 3, observations at x-ray, radio, and infrared wavelengths were uncovering a fascinating universe populated by strange objects, such as supermassive black holes, gamma-ray bursters, quasars, and pulsars. Even the esoteric debates on the past and future of the universe's expansion excited the imagination of the public. In addition, President Kennedy's stated aim of placing an American on the Moon before the end of the 1960s, together with a strong program of planetary exploration, all contributed to a surge in public interest and student enrollment.

The University of California was especially hard pressed. The university was spread over ten campuses, of which four had burgeoning astronomy departments. Students and faculty alike were competing for time on the 120-inch (3-meter) at the university's Lick Observatory. At the same time, the lights of San Jose and its suburbs had begun to interfere with observations at Lick. So beginning around 1965, astronomers at Lick Observatory began to discuss building a second 120-inch telescope at a darker site.

A site committee was appointed, with Robert Kraft of Lick as chair, to investigate a number of potential locations. Merle Walker was their man in the field. He would travel to a potential site, set up his 15-cm telescope, and photograph star trails around the celestial pole to evaluate the seeing (see note 4.8). Among the places he tested were Kitt Peak and Flagstaff in Arizona, Junipero Serra in California, Mexico's Baja California, and Cerro Tololo in Chile. In his 1971 report he summarized some general conditions required for good seeing and gave the highest marks to Tololo and Junipero Serra. He also suggested that high islands in warm oceans, like Hawaii and the Canaries, might have good seeing. (How right he was!)

Joseph Wampler, a Lick astronomer who was highly respected as an experimentalist, argued strongly for Junipero Serra, a mountain south of Monterey. However, when the Lick astronomers began to explore the possibility of an observatory there, they ran into fierce opposition from the Native American population. Worse, they couldn't find a source of money. Nobody wanted to fund just another medium-sized telescope. The Hale 200-inch had set the mark to beat. If Lick couldn't find a way to build something larger, it would never happen.

Only one telescope larger than the Palomar giant was under construction. The Soviets had been struggling for years to build the 6-meter Big Alt-Azimuth Telescope at the Special Astrophysical Observatory in the Caucasus (fig. 6.1). When it was finally finished in 1976, this monster was 26 meters long and was housed in a dome 58 meters high. It was the first large optical telescope to use

Fig. 6.1. The Soviet 6-meter Big Alt-Azimuth Telescope

the alt-azimuth mount. To eliminate the rotation of the image caused by the mount, two guide stars on opposite sides of the field of view had to be tracked. The system worked, but only marginally. What was worse, the mirror never formed sharp images because of inadequate thermal controls for the mirror and the dome. The telescope had been built largely as a display of Soviet superiority and never produced much in the way of research. No one knows how much it cost.

Between 1970 and 1975, astronomers from the University of California (UC) campuses met occasionally to discuss their best strategy. They had a long list of difficult issues. Foremost among them was the question of how large a mirror to aim for. Clearly, the bigger the better, but a rule of thumb in the telescope game was to build up in steps no larger than a factor of two. Hale's experience was a good guide. When he began building reflectors, the Lick 36-inch (0.9-meter) was the benchmark to beat. So he built the 60-inch (1.5-meter), then the 100-inch (2.5-meter), and finally the 200-inch (5-meter). The UC astronomers eventually settled on a 10-meter mirror as a possible goal.

But how would one build a 10-meter? If one simply copied the design of the Palomar telescope and doubled the size of everything, the cost could easily exceed $1 billion. Even if such a colossus could be built, who would provide the money? What's more, the technology of the Palomar telescope dated to 1948. Surely there was some way to utilize new techniques to build a cheaper telescope. The Lick astronomers needed some imaginative ideas.

Joe Wampler came up with a clever scheme for fabricating the mirror, the key to the project. It should be possible, he said, to cast a monolithic mirror 10 meters in diameter and only a few centimeters thick (note 6.1) He would support it with a complex mechanical framework, to keep it from flexing as the telescope tracked a target. A thin mirror would be lighter and easier to cast than a thick one, and it would cost less. The telescope mount could also be lighter, with more savings in time and cost. Wampler began to design the optics and mount for a 10-meter telescope, using Lick engineers and the Lick shops.

Then, in 1977, a young physicist appeared with a wild idea for a telescope.

THE MAVERICK

As a student, Jerry Earl Nelson (fig. 6.2) had no ambitions to build a large telescope. He was a gadgeteer, interested in all of science but mostly in physics. He enrolled at Caltech, obtained a degree in physics in 1965, and went on for a doctorate at the University of California at Berkeley. There he studied elementary particle physics and did some experimental research for his dissertation.

Nelson gradually lost interest in particle physics after participating in several esoteric experiments. He was drawn instead into high-energy astronomy. In 1970 he began to study the optical pulsation of radio and x-ray pulsars with a number of collaborators, including Joe Wampler. First they examined the pulsar in the Crab Nebula, then Hercules X-1, and later many others. Nelson observed optical pulsations with the 36-inch (0.9-meter) Crossley reflector at Lick Observatory and became familiar with other telescopes of moderate size. He also became acquainted with astronomers throughout the California university system.

In 1975 Nelson joined the Lawrence Radiation Laboratory in Berkeley. Ernest Lawrence, the inventor of the cyclotron and a Nobelist, founded this lab in 1931. It became famous for its creative research in particle physics and as the home of nine Nobel Prize winners. It evolved into a National Laboratory, with a broad mandate for research in controlled nuclear fusion, biophysics, en-

Fig. 6.2. Jerry Nelson, appropriately
wearing a Hawaiian shirt

vironmental science, and the physics of materials. The place was seething with
bright researchers.

Despite his shift of career, Nelson kept in touch with the elementary particle physicists at the lab. Among them was Terry Mast, a young experimental
physicist who at the time he met Nelson was studying the decay of cosmic ray
particles and the physics of particle acceleration. Mast bubbled over with ideas.
The two men became close friends.

Nelson was happily engaged in his pulsar research, turning out publications
each year and gradually building a reputation. Then Luis Alvarez, winner of a
Nobel Prize in particle physics, changed Nelson's life (note 6.2). Alvarez had
been Nelson's mentor at Berkeley and appreciated his talent for creative instrumentation. Nelson had heard about the Lick astronomers' discussions and
was wildly excited about a radical type of telescope that could be built as large
as one liked. After listening to Nelson's arguments, Alvarez called Donald Osterbrock, then the director of Lick Observatory, and recommended that the
astronomers give Nelson a hearing.

Nelson was more familiar with cutting-edge technology than most astronomers. He had kept in touch with experimental physicists like Terry Mast who

were using sophisticated digital electronics, solid-state sensors, and computer-controlled mechanical devices in their work. Nelson realized that, using such technology, it might be possible to fabricate a large mirror in segments that fit together like the pieces of a jigsaw puzzle. Each segment would be easier to make than the whole 10-meter mirror, and if it turned out badly, it would be cheaper to replace.

Each segment could be as thin as a few centimeters if it was supported and aligned with a system of sensors and actuators on its back face. That would greatly reduce the weight of the whole mirror and therefore the weight of the telescope mount. In addition, a thin segment could adjust far more rapidly to temperature changes than a thick mirror. That would help to ensure a sharp focus at all times. Of course, the segments would have to be kept aligned to within a fraction of the wavelength of light (say, one-millionth of a centimeter), but that might be possible with the new technology. In one of Nelson's early designs, sixty hexagonal segments were nested like the chambers of a honeycomb. Such a mirror would be only one-quarter the weight of a conventional mirror of the same size.

Nelson's mirror had another potential advantage. Most large telescope mirrors built up to that date had relatively long focal lengths compared to their diameters. The 200-inch mirror, for example, has a focal ratio (focal length divided by mirror diameter) of f/3.3. The focal ratio determines how "fast" the mirror is—that is, how much it concentrates the light it receives onto cameras and spectrographs at the prime focus. The smaller the ratio, the shorter the photographic exposures could be. A segmented mirror could, in principle, have a focal ratio as small as f/2. And as a bonus, its shorter focal length would allow a shorter mount and therefore a smaller dome, with cost savings all around.

When Nelson first outlined his scheme, it sounded like pure fantasy to many of the UC astronomers. One might be able, conceivably, to hold such a gaggle of glass pieces together with some high-tech gadgetry, but how could one possibly grind and polish each piece so that it would fit precisely into the paraboloidal shape of a telescope mirror?

Most mirrors are first ground to a spherical shape, because that is the easiest to obtain by standard methods. Then a tedious process of *figuring* follows, in which tiny amounts of glass are polished off to obtain the desired parabolic shape. In a single monolithic mirror, this figuring process is simplified by the fact that the mirror has cylindrical symmetry. A polishing tool can be set at a particular radius, and as the mirror rotates, the tool removes the same amount of glass all around a circle.

In the segmented mirror that Nelson proposed, each segment would have to have a slightly different nonspherical shape. The optical industry had been producing aspheric lenses for microscopes for a long time and had some experience with off-axis telescope mirrors, but Nelson's proposed segments would challenge the best of opticians. The cost of producing a 10-meter mirror in segments could be outrageous if one had to develop a whole new technology.

Nelson had to find a better way. If only a spherical mirror would satisfy the requirements of the astronomers, it would be so much easier to fabricate in segments. But a spherical mirror is plagued with *spherical aberration,* which means that light reflected from different parts of the mirror focus at different points along the axis. The result is a blurred image, and that is why figuring to a parabola is necessary. Most Lick astronomers were skeptical about Nelson's scheme but felt there was no harm in his pursuing it if he could find the time and money. Meanwhile, they encouraged Wampler to continue designing a telescope with a monolithic mirror.

BRAINSTORMING

Nelson knew he would have to prove that his ideas could work. But he was no expert on mirror fabrication, so he looked around for advice.

Fortunately, the ideal partner was already working at the Lawrence Berkeley Laboratory. Jacob Lubliner was a professor at the University of California at Berkeley and a senior researcher at the lab. He was an expert on the mechanics of fluids and solids, familiar with the mathematics of stresses and strains. He had interests that spanned such diverse subjects as wave propagation in solids, the flow of blood in veins, the flagella of microbes, and the cracking of concrete. He was the very man Nelson needed.

In lively discussions, they hammered out a clever scheme. Perhaps there was a way to retain the simplicity of polishing to a sphere while still approaching the desired parabolic shape. Suppose you bent the glass blank slightly before you began to polish? If the glass was strong enough not to crack, it might spring into the correct shape once it was released.

There was some precedent for this scheme. Bernhard Schmidt, a German optician, had bent a glass blank in order to grind a circular correcting lens for a wide-field reflecting telescope (see chapter 2). The now-famous Schmidt telescope design is used for compact amateur telescopes as well as the professional 48-inch (1.2-meter) at Palomar.

The idea of warping a spherical mirror into a permanent shape was also not new, but nobody had tried to *grind* a mirror while it was bent and then allow it to relax into a new shape. The scheme would work only if the glass was perfectly elastic. Any "memory" of its bent form could distort the desired shape. Fortunately, Lubliner knew already that many glasses remain elastic right up to the breaking point.

The problem of how much to bend the blank was solvable, in principle. Lubliner had the mathematical tools to predict the forces that would have to be applied to a blank. Nelson had to work out the apparatus that could do the job. Of course, the original thickness that is used to predict the required forces changes as a result of grinding, and these changes had to be predicted, too. The real test would come when a polished circular blank was cut into a hexagon. Would cutting the segment deform it enough to make it useless? Nelson and Lubliner would have to cross that bridge later on.

TRIALS AND TRIBULATIONS

Over the next two years, Nelson and Lubliner tested every aspect of the idea.

In Nelson's preliminary design, the mirror consisted of sixty segments. Each would have the same hexagonal shape, about 1.4 meters in width and about 10 cm thick. The complete mirror would have a very short focal ratio, f/2, which would allow shorter photographic exposures, a shorter mount, and a smaller dome, all desirable qualities.

Lubliner and Nelson worked out the pattern of forces that would have to be applied to a circular blank of glass so that after a spherical surface had been ground and the forces removed, the blank would spring into the desired shape, a segment of a paraboloid. To test the idea, they decided to build a quarter-scale model of a segment of the proposed 10-meter mirror. They chose a circular glass blank 36 cm in diameter and 2.5 cm thick, the same proportions as a 1.4-meter segment 10 cm thick.

To bend the blank, they applied a uniform pressure with a rubber pad across the back of the blank as well as shear forces and torques at the rim of the blank.

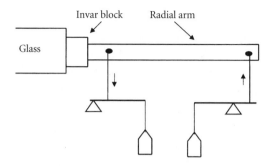

Fig. 6.3. The weights and levers used to stress a blank while being polished to a spherical shape

A simple jig of weights and levers was built to apply the forces (fig. 6.3). They epoxied blocks of invar, a low-expansion steel, at twenty-four places around the rim of the blank. A radial arm was attached to each block, from which weights could be hung.

Surprisingly, the rubber pad turned out to be the most difficult piece of the setup. Small nonuniform patches in a pad could cause unwanted distortions in the blank. Lubliner and Nelson experimented with a bag of water and with a variety of elastic pads made of neoprene and natural rubber. With each choice they measured the size of the distortions the pad introduced, using standard interferometric methods. After much experimentation they settled on a three-ply natural rubber pad.

When they felt that they had developed the best system for bending a blank, they employed the commercial firm of Tinsley Laboratories to grind and polish it. A second blank, made of Cervit, a low-expansion ceramic, was also prepared for comparison. At every stage of the process, exhaustive tests were made of the accuracy and reproducibility of the predicted shapes.

In 1980, after two years of experimentation, Nelson and Lubliner decided that the stress-polishing technique could work satisfactorily. With only two polishing operations, they could produce a surface accurate to within 0.03 micron, or about one-twentieth of a wavelength. They published a description of the method and the results in two articles.

JUGGLING A STACK OF DISHES

At the same time, Nelson was working with Terry Mast and an engineer, George Gabor, on the support and alignment control systems. Each segment would have a passive mechanical system to prevent it from sagging under the force of gravity. Three so-called whiffletrees (fig. 6.4) would support the weight of the segment at thirty-six points and spread the load uniformly. In addition, an active control system was provided to keep all the segments in alignment. Two sensors would be mounted on each intersegment edge, for a total of 168. Each sensor is a capacitor, consisting of two parallel metal plates. Any displacement of a segment would change the spacing of the plates and hence its capacitance, which could be measured extremely accurately.

The sensors would compare the position of an edge to a corresponding edge on a neighboring segment. A difference as small as half a millionth of a centimeter could be measured. Three mechanical actuators on each segment could tilt it in two directions at right angles (tip and tilt) as well as up and down. A special microcomputer would compare the data from the 168 sensors and order the actuators to correct the positional errors. The process could be repeated several times a second to keep the segments aligned despite the forces of gravity, wind, and thermal expansion.

Nelson and Mast decided to publicize their concepts so as to get some reactions from other physicists and engineers. In December 1977 Nelson presented his basic idea of a segmented 10-meter mirror at a conference on optical telescopes of the future, in Geneva, Switzerland. At this stage Nelson was proposing a mirror with fifty-four segments around a 3.6-meter central mirror. He got some valuable feedback and no insurmountable objections. By January 1979 he and Jacob Lubliner had progressed far enough with their tests of stress polishing to be confident that it would work. So Nelson presented the whole scheme at a seminar on astronomical instrumentation, in Tucson, Ari-

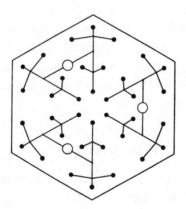

Fig. 6.4. Three whiffletrees *(shown in solid lines)* each support a fraction of the load of a segment on twelve points. In this sketch the dark dots are the support points, and the open circles are the actuators that control the tip and tilt of the segment.

zona. Mast held forth at several instrumentation conferences as well, and they both published preliminary reports on their work. The favorable comments they received emboldened Nelson to submit a detailed proposal to the UC astronomers.

Don Osterbrock, director of Lick, formed a committee (the "Graybeards") to choose between Wampler's monolith and Nelson's segmented mirror. The committee included two astronomers from each of the faculties at Berkeley, Los Angeles, Santa Cruz (Lick), and San Diego. Len Kuhi and Charles Townes (the inventor of the laser) represented Berkeley, Margaret Burbidge and Harding Smith came from San Diego, while Roger Ulrich and Holland Ford came from Los Angeles. Don Osterbrock and Robert Kraft represented Santa Cruz. Several other astronomers sat in to advise, including Joe Wampler and Sandra Faber from Lick and Ivan King from Berkeley. All of these people were users of the Lick Observatory, a criterion that Osterbrock insisted on.

At a meeting at the Berkeley Faculty Club on November 27, 1979, the Graybeard Committee heard presentations by Wampler and by Nelson. Each scheme had its merits and its risks, and the committee was split. In the end they voted five to three to accept Nelson's design provisionally, and they encouraged him to proceed further. He would have to construct a Technical Demonstration, however, a prototype of a complete system for one segment that would test all aspects of the proposed system.

At this point a substantial sum of money was needed to develop the design of the telescope from a concept to a complete working instrument. The Lick astronomers had kept Robert Sinsheimer, the chancellor at UC Santa Cruz, advised about the project. Sinsheimer took the funding problem to David Saxon, president of the UC system, who convened a council of the chancellors of the four campuses with astronomy programs. This group was split four ways about whether to devote university money to the project, but Saxon persuaded them to proceed. He would find the million dollars a year needed to fund Nelson's Technical Demonstration and to design the rest of the telescope hardware.

The Technical Demonstration had several specific goals. First, a thin stressed circular blank had to be polished to a spherical surface with a prescribed radius of curvature. Then the blank had to be cut to its final hexagonal shape, and the distortions introduced, if any, had to be measured and corrected. Next, the active and passive systems for supporting the segment had to be built and tested. All this would require constructing the whole system of actuators, sensors, and computer software. A target date for completion of the demonstration was set for the end of 1983.

Nelson wrestled with the optimum size of a segment. The smaller the segment, the easier it would be to figure but the more segments, with all their attendant hardware, the final mirror would contain. It was a tradeoff between cost and feasibility. In the original sixty-segment design, a central segment 3.6 meters across was included to act as a test plate for the alignment of all the other segments. Eventually Nelson and Mast worked out a better scheme for aligning the segments that didn't require this large central segment. They were able to reduce the number of segments to thirty-six, all identical at 1.8 meters across and 7.5 cm thick (fig. 6.5).

By early 1983 a 1.8-meter segment had been successfully polished to the specified 10-meter radius of curvature. The surface quality was good enough to ensure that an image of a star would lie in a circle of only 0.2 arcsecond, which is five times smaller than the image blur usually produced by the atmosphere. Nelson was delighted. (An arcsecond is 1/3600 of a degree of arc, the angle a pinhead makes at a distance of 200 meters.)

Of course the image quality of the final mirror would depend on the aberrations introduced by misalignments of all thirty-six segments, and although

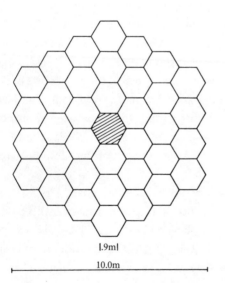

l.9ml

10.0m

Fig. 6.5. As finally configured, the Keck mirror consists of thirty-six segments arranged in a honeycomb pattern. The central segment is omitted to allow the light beam to reach the Cassegrain focus behind the mirror.

that problem could be studied mathematically, it would require rigorous testing, too.

With the Technical Demonstration progressing nicely, the Lick astronomers returned to the question of a site for the 10-meter telescope. Everyone was hearing good reports about the observing conditions at Mauna Kea. The seeing was good, the sky was dark, and nighttime clouds were rare. As of 1979, seven telescopes with mirrors 3 meters and larger were showing beautiful results, and other telescopes were on the way. The road to the summit had been improved, permanent power lines were in, and a midlevel dormitory had been built. The UC astronomers were impressed and decided that Mauna Kea was their first choice for an observatory.

Robert Kraft, by now the director of Lick Observatory, negotiated with John Jefferies, director of the Institute of Astronomy in Hawaii, for a site near the summit of the great volcano. After considerable diplomatic maneuvering with environmentalists and representatives of the native Hawaiians, permission was granted. All the Californians needed now was the money to build their telescope.

THE CALIFORNIA TEN-METER FINDS
A PARTNER AND A SPONSOR

The price tag on Nelson's design for a 10-meter was roughly $100 million. It was comparable in cost to the 200-inch (taking inflation into account) but offered four times the collecting area. Nevertheless, it was a lot of money. With the cost of all the auxiliary equipment, the total might reach $150 million. Where could anyone find a sum like that?

The National Science Foundation was the natural source to consider. But the NSF doesn't grant funds of this size without an exhaustive evaluation by a committee of experts. Despite Nelson's Technical Demonstration, many experts still regarded a large segmented mirror as a risky proposition. Moreover, the NSF also imposes a rigid requirement: a facility built with public funds must be available to the public. That meant that the astronomers of the University of California would have to share the 10-meter with any professional astronomer who could propose a reasonable observing program. But by this time about a hundred astronomers and students in the UC system needed telescope time. They felt they had to have their own telescope.

Accordingly, David Saxon and Robert Sinsheimer began to look for private funds. Through a roundabout route they heard of a possible donor. Max Hoffman, a wealthy San Francisco businessman, had died in 1981, leaving a substantial fortune to his widow, Marion. She, it turned out, was receptive to the idea of providing about half the funds ($36 million) for an observatory in his name. The legal documents were typed and prepared for her signature, but she died a few days before the deal could be closed. Her estate went to probate court, and despite the opposition of the trustees of the new Hoffman Foundation, the UC was awarded the full $36 million. So far, so good.

Now the problem was to find the other half of the construction costs. Sinsheimer convinced the Lick astronomers that they would have to share their telescope, and he approached Marvin Goldberger, president of Caltech, as a possible partner. Goldberger, a physicist, was immediately intrigued by the novel technology. He spoke with the Caltech astronomers, who were less than overwhelmed by the prospect. Nevertheless, Goldberger and David Gardner, the new president of the UC system, somehow pushed through an agreement. Caltech would find the remaining half of the construction costs.

Goldberger began to look for million-dollar donors. He didn't have far to look. Howard Keck, one of the trustees of Caltech, was also the chairman of the William M. Keck Foundation, one of the largest philanthropic institutions in the country. As a wildcat operator, William Keck (Howard's father) had struck oil in Coalinga, California, in 1922. He later founded the Superior Oil Corporation and served as its chairman until the age of eighty-three. In 1954 he established a nonprofit foundation with his substantial fortune, with the aim of funding worthy programs in science, engineering, and medicine. The Keck Institute for Neurosciences, for example, is one of many research and development organizations that benefited from Keck's philanthropy. When Keck died in 1964, his son Howard took charge of the foundation, which prospered mightily. Its endowment grew to over $1.5 billion, despite paying out some $875 million in grants.

Goldberger convinced Howard Keck of the scientific potential of this 10-meter telescope. With Keck's enthusiastic support, the Keck Foundation granted $72 million, the whole cost of the telescope, with the condition that the observatory be named for William M. Keck.

That brought an embarrassment of riches, for now the UC had the problem of what to do with the Hoffman donation. A bright idea surfaced: why not build a second telescope, identical to the first, with Hoffman's $36 million? Two for nearly the price of one. Unfortunately, the trustees of the Hoffman Foundation didn't want to share the glory of a world-class telescope with another donor. They withdrew their approval, and the UC, with regret you may be sure, returned the grant of $36 million. But the 10-meter telescope was now well under way.

The University of California and Caltech formed the California Association for Research in Astronomy (CARA), a nonprofit corporation for building and operating the telescope. Caltech and the UC would each have 40 percent of the telescope time, the University of Hawaii would have 10 percent, and the remaining 10 percent would be allocated to engineering and maintenance. The UC agreed to pay the operating costs of the observatory for twenty-five years, as its share of the total.

CONSTRUCTION

On September 12, 1985, a group of dignitaries held a groundbreaking ceremony at the top of one of the great cinder cones near the summit of Mauna Kea. From this superb site the huge mirror would have a marvelous view of the black nighttime sky.

The Keck telescope is designed as a Ritchey-Chrétien Cassegrain, with a prime focus, Cassegrain focus, and two Nasmyth foci (note 6.3). The primary mirror has the shape of a concave hyperboloid, not the usual paraboloid, and the secondary mirror is a convex hyperboloid (note 6.4). This circular secondary mirror, 1.45 meters in diameter, proved to be terribly difficult to shape. David Hilyard and Joe Miller worked long and hard in the Lick optics shop to perfect its figure.

Jerry Nelson decided to make the mirror segments out of Zerodur, a mixture of crystalline and vitreous forms of glass that the Schott Glaswerke produces in Mainz, Germany. This substance expands and contracts extremely little when the air temperature changes, which greatly eases the problem of keeping the mirror in shape. Schott machined Zerodur blanks into circular, meniscus-shaped disks and shipped them to Itek Optical Systems in Lexington, Massachusetts. Itek used the stress-polishing technique to shape the disks. The final cutting and testing was also done there.

The segmented mirror has the very fast focal ratio of $f/1.75$ at prime focus.

At the Ritchey-Chrétien Cassegrain focus the focal ratio is a more modest f/15. The mirror weighs only 14 tons, comparable to the 200-inch but with four times the light-gathering power.

The telescope mount is an alt-azimuth and tracks a target with the aid of a computer. Altogether the telescope and its mount weigh only 298 tons, about half as much as the 200-inch. The Keck's dome (or *enclosure*) received special attention. Its designers at Coast Steel Fabricators were well aware that hot air rising out of a dome could easily ruin the seeing. So they eliminated as many sources of heat inside the dome as possible. They also provided heavy layers of insulation and an air-conditioning system to keep the dome cool during the day, as well as powerful fans to circulate air at night. The dome has a squat shape, wider than high. It is 30 meters tall, weighs 700 tons, and could easily fit inside the dome of the 200-inch.

Construction of the Keck Observatory proceeded at a brisk pace. The foundations for the telescope were completed by 1987, and the dome was erected the following year. Meanwhile, the mount's vertical tube was completed in Spain and tested for flexure. By 1989 the telescope mount was in place. Next the mirror cell, with its complex dual support system, was attached. On November 20, 1990, the first nine mirror segments were installed and tested. Additional segments were added over the next year, and on April 14, 1992, first light was achieved with all thirty-six segments. The optical performance of the telescope surpassed all expectations.

Mauna Kea is an excellent place to view the sky, but a difficult place to work. At nearly 4,300 meters, the lack of oxygen makes thinking a struggle. So CARA decided that it needed a headquarters on the island where visiting scientists and technicians could work comfortably. A site was chosen in the quaint village of Waimea, at the modest altitude of 760 meters. In May 1988 ground was broken there for an office building and shops. The W. M. Keck Observatory building was completed a year later. With a dedicated microwave link to the summit, an astronomer would be able to see her targets and operate the telescope remotely from there.

KECK II

Everyone knows that building copies of a proven design is much easier than building the prototype. Following this principle, CARA embarked on construction of a second Keck telescope just a few hundred meters from the first, with an additional grant of $70 million from the Keck Foundation. The ultimate plan, which we will describe later, was to couple the two 10-meter telescopes, together with a few smaller ones, to form an optical interferometer. This would allow studies of very faint objects at high angular resolution, the ultimate dream of a nighttime astronomer.

Ground was broken for Keck II on November 7, 1991, five months before Keck I met its full performance test, and the work was completed in March 1995. Figure 6.6 shows a schematic of the twin telescopes, and figure 6.7 is a panoramic view of the summit of Mauna Kea, with the twin domes in place.

The Californians deserve much credit for managing such a complex project in such a hostile physical environment and in such a short time. The key person in this endeavor, aside from Jerry Nelson, was the project manager, Gerald Smith. Smith had extensive experience as a manager at NASA. He was responsible for the construction of the Infrared Astronomical Satellite and

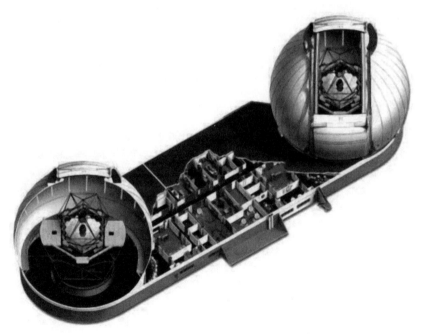

Fig. 6.6. A schematic of the two Keck telescopes in their enclosures and the laboratory building that joins them

Fig. 6.7. The Mauna Kea Observatories. In the foreground, from left to right, are the UH 0.6-meter telescope (small white dome), the United Kingdom Infra Red Telescope, the UH 2.2-meter telescope, the Gemini North 8-meter telescope, and the Canada-France-Hawaii Telescope. On the right are the NASA Infrared Telescope Facility and the twin domes of the W. M. Keck Observatory; behind is the Subaru Telescope. In the valley below are the Caltech Submillimeter Observatory (silver) and the James Clerk Maxwell Telescope (white).

NASA's 3-meter on Mauna Kea. He joined the team in 1985 and retired after Keck II was completed. Hans Boesgaard was the man on the site. Originally chief engineer for Hawaii's Institute for Astronomy, Boesgaard supervised the contractors during the construction of the Keck telescopes and eventually became the manager of the facility.

THE INSTRUMENTS

A big mirror, by itself, does not make an observatory. The potential for innovative research depends as well on the focal plane instruments that detect and analyze the light. Keck I and II are equipped with an impressive arsenal of specialized instruments. Here is a sample of their inventory.

The Near Infrared Camera (NIRC) is used to make digital images of extended objects, such as galaxies or planets, at wavelengths between 1 and 5 microns. It achieves diffraction-limited resolution, 0.02 arcsecond, at 1 micron (see note 5.2 on the diffraction limits of telescopes). Cooled down to 77 kelvin, the device is so sensitive that it could detect a candle's flame on the Moon. A second-generation version, NIRC 2, has been installed recently at one of the Nasmyth foci. At its heart lies the detector, a 1,024 × 1,024 pixel indium-antimonide array.

If you have a near infrared camera, you also need one for the far infrared. The Long Wavelength Infrared Camera creates diffraction-limited images in the band from 8 to 15 microns. It is most useful for penetrating very dusty regions, such as the cores of galaxies and the gas clouds where hot stars are born.

Often it is necessary to obtain the complete spectrum of a faint object. The Echellette Imager and Spectrograph does just that. Its 2,048 × 4,096 pixel array captures everything from the ultraviolet to the near infrared in one long exposure.

The Keck telescopes are renowned for their spectroscopy of faint objects such as quasars. The Near Infrared Spectrometer is a key instrument for this purpose. It features an infrared *echelle grating* (note 6.5) for the 1 to 5 micron region, with another 1,024 × 1,024 pixel indium-antimonide array detector.

DEIMOS, the Deep Extragalactic Infrared Multi-Object Spectrograph, is designed to observe up to seventy-five distant objects at once and to obtain all their spectra simultaneously. It has an enormous array detector with 8,192 × 8,192 pixels. Its first task will be to survey some fifteen thousand distant galaxies and help to create a three-dimensional model of their distribution in space.

SOME RECENT SCIENCE HIGHLIGHTS

Even a short list of discoveries made with the Keck telescopes could take many pages. Just to get a flavor of the science that is coming out of the observatory, we'll touch on a few general areas.

One of the big questions in astronomy (and one that fascinates the public) is whether we are alone in the universe. Is there any place out there where life, even bacterial life, could exist? To find out, astronomers are using the Keck telescopes, among others, to search for extrasolar planets. And they're finding them, by the dozen.

Steven Vogt (UC Santa Cruz), Geoffrey Marcy (UC Berkeley), and R. Paul

Butler (Carnegie Institution) are among the most successful of these explorers. They have spent three years so far searching the nearest one thousand Sun-like stars for planets. They used a high-resolution spectrograph, first at the Lick 24-inch (0.6-meter), then at the 120-inch (3-meter), and finally at the Keck. Their trick is to look for periodic Doppler shifts in the spectrum of a single star. Such shifts reveal the presence of an unseen companion, a planet, whose gravitational pull causes the star to wobble. From the details of the signal, they can derive the orbit and mass of the planet.

In November 1999, for example, they found six Jupiter-sized planets, bringing their total to twenty-eight. Five of these lie in the so-called habitable zone around the parent stars, where the temperature is temperate enough for life as we know it to exist. Oddly enough, some of these planets revolve around their star in days, not years.

These same astronomers found two planets in January 2001 that are smaller than Saturn, perhaps three times smaller than Jupiter. But so far their technique cannot detect Earth-sized planets. Something like a hundred extraterrestrial planets are known by now.

Another exciting research area concerns the enigmatic gamma-ray bursters. These strange transient events were detected in the 1960s by the VELA satellites, which were monitoring Soviet compliance with a ban on nuclear tests in space. For a long while the location of the bursters was uncertain, but the fact that they scatter evenly over the sky suggested that they lie outside our galaxy.

The burst of gamma rays (extremely short wavelength x-rays) can last as little as a few milliseconds or as long as a few minutes. Because the burst is so brief, and because exact positions are difficult to fix at gamma-ray wavelengths, the cause of these strange events remained uncertain for decades. Finally, in May 1997, the Italian-Dutch BeppoSAX satellite got a good position for a burster. Astronomers at Palomar discovered an optical *afterglow* in a host galaxy at that position. Then, another group used the Keck I and II spectrographs to obtain the galaxy's redshift, $z = 0.8$ (z equals the wavelength shift of a spectral line divided by its laboratory wavelength). That placed the burster halfway across the observable universe.

Since 1997, many more afterglows have been detected. Spectroscopy of the glows with the Keck telescopes has indicated that the burst is the sign of a supernova, the explosion of a star with a mass twenty to thirty times that of our

Sun. From the redshift of the hydrogen Lyman lines, the recession velocity and thence the distance and age of the burster can be determined. One of the most distant bursters was born when the universe was a mere hundred million years old. These incredibly bright objects offer a means to explore the creation of galaxies early in the life of the universe.

Gravitational lensing is another technique that allows astronomers to look at the very early universe. Einstein's theory of general relativity predicts that the gravity of a massive object can deflect the light from a distant source. In fact, the deflection of starlight by the Sun, seen during the total solar eclipse of 1919, was the first confirmation of Einstein's theory. Astronomers have since learned that a nearby galaxy can focus the light from a distant galaxy, just as a lens does. The lensing effect collects more light from the distant source than the observer would otherwise see, so the source is much brighter. Lensing produces a pattern of arcs and spots from the distant galaxy, and if the mass distribution of the foreground galaxy is known, the distance and mass of the farther galaxy can be determined.

In October 2001 Richard Ellis (Caltech) and an international team used the Hubble Space Telescope to discover the lensing pattern (see fig. 5.9) of a very young galaxy at the extreme limits of the universe. Then, applying the enormous light-gathering power of the Keck telescopes, the team was able to obtain a spectrum of the galaxy and to determine its recession velocity, and hence its distance. Its distance corresponds to an age of 13.4 billion years. The Big Bang probably occurred about 14 billion years ago, so this infant galaxy was born only 600,000 years later! Ellis and his team interpret the object as a building block for a normal-sized galaxy. Presumably the early universe was filled with such building blocks. This is a crucial finding, as it bears on the manner in which the galaxies we see today were formed.

For the past decade, two international groups have used a special type of supernova (Type Ia) as a standard candle to search for changes in the rate of expansion of the universe (note 6.6). Most of these supernovas lie at redshifts less than $z = 0.8$, so they are much closer than the spectacular quasars, which have redshifts up to $z = 5$ and 6. Quasars, unfortunately, vary too much in intrinsic brightness to be useful as standard candles. In contrast, the rates of brightening and dimming, together with the associated changes in color, are good indicators of the absolute brightness of these supernovas.

The Super Nova Cosmology Project is led by Saul Perlmutter (Lawrence

Berkeley National Laboratory), and the High-Z Supernova Search is led by Brian Schmidt (Mount Stromlo Siding Springs Observatories). Both groups have used lots of time on 4-meter telescopes, such as the Blanco at Cerro Tololo, the Canada-France-Hawaii Telescope and the United Kingdom Infra Red Telescope at Mauna Kea, and the La Silla 3.6-meter, to find examples of these supernovas in their host galaxies. To obtain the supernovas' redshifts, and to measure their apparent brightness, these teams resorted to the enormous light-gathering power of the Keck telescopes, the Hubble Space Telescope, and, most recently, the European Southern Observatory's Very Large Telescope.

Both teams have concluded that the average mass density of the universe is too low to cause an ultimate slowing down and collapse, a Big Crunch. In fact, they conclude that the expansion of the universe is *accelerating*, and that requires that some mysterious so-called dark energy to offset gravity's pull. According to their latest estimates, 25 percent of the mass in the universe consists of ordinary and dark matter, and a full 75 percent corresponds to this dark energy. So at this point, astronomers are in the uncomfortable position of not knowing what most of the universe is made of.

The Keck telescopes have also been employed to refine estimates of the famous Hubble constant, to study intergalactic gases, and to discover the most distant quasars. The list of accomplishments could go on and on, but we must move on. Suffice it to say that the Keck telescopes represent the cutting edge of astronomical technology and are, as we shall see, the prototypes of even larger telescopes. Moreover, the telescopes are now equipped with adaptive optics, which remove much of the blurring effects of the Earth's atmosphere.

SPINNING GLASS

WHILE JERRY NELSON was perfecting the segmented mirrors for the Keck telescopes, another quiet revolution in optical science was under way in Tucson, Arizona. Nelson had decided that casting a monolithic mirror much larger than the Palomar 200-inch would be too difficult. Even if you succeeded, you would probably need a system of actuators to prevent it from flexing as the telescope moved. In that case, he concluded, you might just as well aim for a segmented mirror from the outset.

Roger Angel wasn't so sure. He had fresh ideas for casting a large rigid mirror. And he guessed that making small corrections to a stiff mirror should be much easier than keeping a collection of segments aligned.

James Roger Peter Angel, to give him his full name (fig. 7.1), is now a regent professor of astronomy at the University of Arizona, the director of the world-famous Steward Observatory Mirror Lab there, and a recipient of the MacArthur "genius" award. He has pioneered a unique process for the fabrication of large telescope mirrors. But he never expected to end up where he is today.

EARLY DAYS

Angel was born in Lancashire, England, in 1942. A bright lad, he obtained a scholarship at age seventeen to Oxford University, where he studied physics. With his bachelor's degree in hand he migrated to Caltech in 1963 for a master's and then returned to Oxford for a doctorate. When he finished in 1967,

Fig. 7.1. Roger Angel

he was looking for work. He applied for a postdoctoral research position at Columbia University, intending to continue his doctoral research on the polarization of starlight. When he arrived in New York he learned that his intended mentor, Robert Novick, had switched fields and was now studying the polarization of x-rays from astronomical sources. Angel joined in enthusiastically. One of his earliest scientific papers concerned oscillations in the x-rays from Scorpius X-1, a strange binary star system.

In 1975 Angel joined the faculty of the University of Arizona, attracted by the prospect of lots of observing time at the Kitt Peak National Observatory, just an hour's drive from Tucson.

Angel earned his keep by teaching courses in astronomy and optics. He also teamed up with Harold Stockman, Nicholas Woolf, and several other British astronomers at Kitt Peak. Together they exercised Angel's expertise in polarization. They studied the Crab Nebula and active galaxies, and in 1979 they discovered a white dwarf with a magnetic field a hundred million times stronger than that of the Sun. Angel became something of an expert on magnetic white dwarfs and their role in x-ray binaries. Throughout the 1970s he was moving along a conventional track of research and teaching. At the same time, he was becoming known as a talented designer of unconventional instruments.

For example, around 1979 Angel got a bright idea for a most unusual stel-

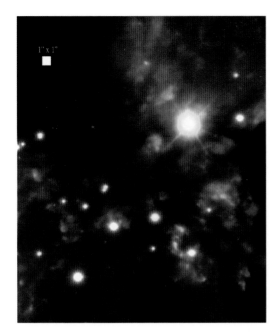

Fig. 3.7. The Becklin-Neugebauer cloud (the bright starlike object surrounded by wispy clouds) and the Kleinmann-Low nebula lie within the Orion Nebula. Shock-excited molecular hydrogen emits a line at 2.12 microns (rendered blue in the image), and ionized hydrogen emits a band around 2.2 microns (rendered in red).

Fig. 5.4. Massive stars in 30 Doradus, a nebula in the Large Magellanic Cloud, generate powerful winds that compress the surrounding interstellar clouds. As a result, new stars are born. NICMOS observations at infrared wavelengths pierce the dense dust clouds to reveal this star nursery.

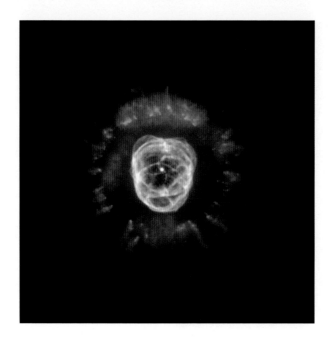

Fig. 5.6. The Eskimo Nebula, NGC 2237, as seen by the Wide Field Camera on the Hubble Space Telescope

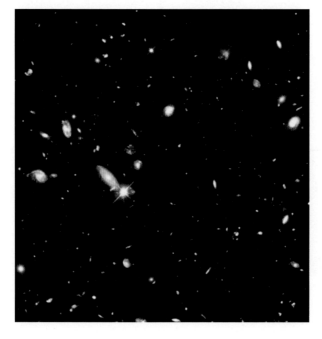

Fig. 5.7. The Hubble Deep Field, an image of fifteen hundred distant galaxies in a narrow tunnel through space

Fig. 8.3. A highly resolved infrared view of the center of the Milky Way, obtained in 2002 with Yepun, one of the units of the Very Large Telescope. The length of the bar indicates a distance of 1 light-year. We can see stars of different ages, some still enveloped in their clouds of dust. The two arrows point to Sagittarius A, the exact center of the Milky Way, where a supermassive black hole resides.

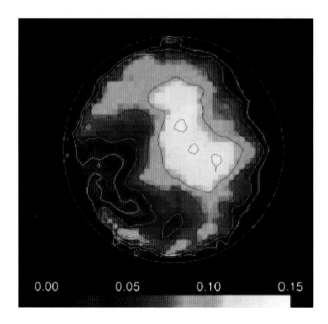

Fig. 9.4. This 2-micron image of Titan, Saturn's largest satellite, was obtained with speckle interferometry at the Keck II telescope in 1999. Titan has an angular diameter of 0.7 arcsecond and a linear diameter of 5,150 km. The Cassini probe passed close to Titan on July 2, 2004, and obtained higher-resolution images of the cloudy surface.

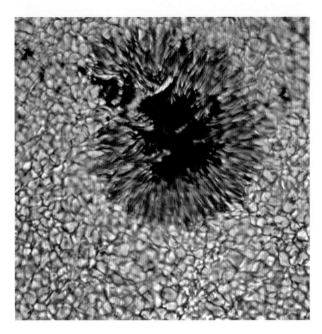

Fig. 9.8. Fibrils in the penumbra of a sunspot are shaped by the magnetic field lines that spread from the center. The sunspot lies in a field of solar granules, each about the size of Alaska. Their sub-arcsecond dark borders can serve in place of a point source for an adaptive optics system.

Fig. 9.11. A 2.2-micron image of Io, one of the four Galilean moons of Jupiter. It was obtained with the adaptive optics system NAOS-CONICA on the Very Large Telescope in December 2001. Io has a diameter of only 1.2 arcseconds, but the system resolves features 0.063 arcsecond in size.

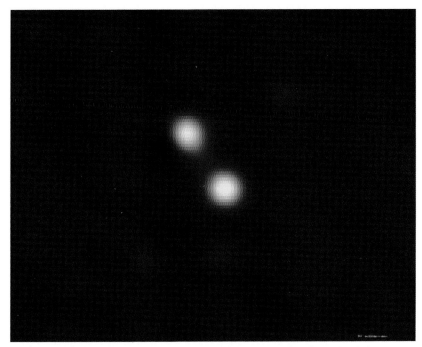

Fig. 10.8. The first aperture synthesis image of a close binary, Capella, obtained with the interferometer of the Cambridge Optical Aperture Synthesis Telescope. The separation of the stars is about 50 milli-arcseconds.

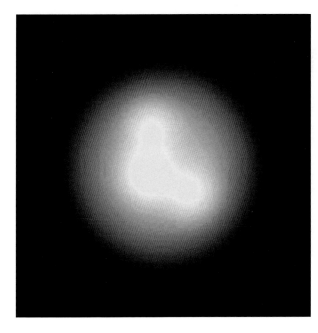

Fig. 10.9. Bright spots on the surface of Betelgeuse, at a wavelength of 700 nm, obtained with the COAST interferometer

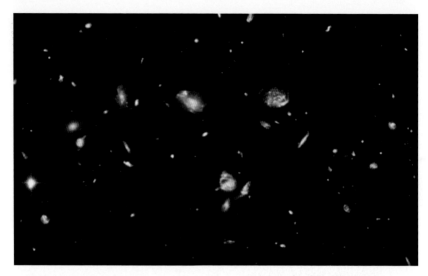

Fig. 11.1. Part of the Hubble Ultra Deep Field. The image shows galaxies from the present back to when the universe was less than a billion years old.

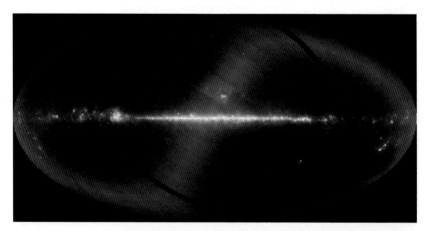

Fig. 12.1. The whole sky as imaged at 12, 60, and 100 microns by the Infrared Astronomical Satellite. In this composite picture blue indicates hotter regions, red cooler regions. The horizontal band is the plane of the Milky Way. The dark curved feature is produced by warm dust in the plane of the solar system.

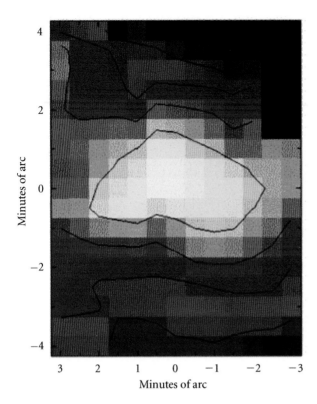

Fig. 12.2. The Infrared Space Observatory captured the first image of a dense core of gas molecules and dust in which a star will form. This picture was obtained at a wavelength of 160 microns.

Fig. 12.4. One of the first images from the Spitzer Space Telescope shows this stellar nursery in the Large Magellanic Cloud. An expanding supernova remnant *(upper part of the image)* has triggered the birth of stars in this nebula, Henize 206.

Fig. 12.6. Our Milky Way galaxy apparently has a "bar" at its center, similar to the one in Messier 83. This galaxy is located in the southern constellation Hydra.

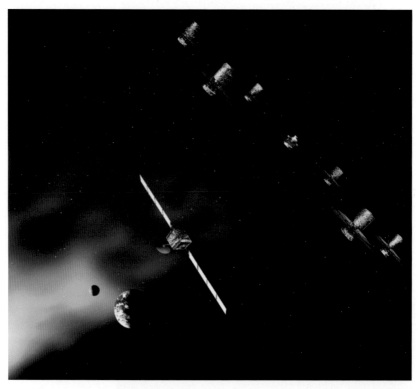

Fig. 12.7. An artist's conception of an infrared nulling interferometer called Darwin with its fleet of satellites flying in formation

lar spectrograph. All spectrographs at that time could record only one star at a time. An astronomer would set the image of the star on the slit of the spectrograph, expose his film for a predetermined time, and then move on to the next star. It was a slow and tedious procedure, wasteful of precious telescope time. The light from all the stars in the field of view was being wasted while a single star was on the spectrograph slit. What if that wasted starlight could be captured? That would multiply an astronomer's effective observing time by a huge factor.

The key to Angel's invention was optical fibers, which are able to pipe light around corners (note 7.1). These wispy filaments of glass had become readily available. Angel conceived the idea of constructing a metal template, designed for a specific region of the sky. The template would lie at the focal plane of a telescope. It would have a number of holes drilled at the precise locations of bright stars in the focal plane, and an optical fiber would transmit the starlight to a position along the length of a standard spectrograph's slit. In effect, Angel would parallel-process the field of stars. This multiple-object spectrograph (nicknamed Medusa for its bundle of fibers) worked brilliantly and yielded at least a tenfold increase in an astronomer's productivity.

In the early 1970s Angel was part of a team building an unconventional telescope. The University of Arizona had joined the Smithsonian Institution in building the so-called Multi-Mirror Telescope on Mount Hopkins, Arizona. The MMT consisted of a battery of six 1.8-meter mirrors on an alt-azimuth mount (see fig. 4.10). The light from all six mirrors was piped into a common focal plane. That made the MMT equivalent to a single 4.4-meter telescope, the third-largest telescope in the world. And in principle the mirrors could be combined to form an optical interferometer. The MMT eventually produced a lot of good science, but it was finicky and had many teething problems. A single 4.4-meter mirror would have been preferable, although expensive and difficult to produce with the existing technology. Angel began to think of new ways to create large mirrors.

SPIN CASTING

The ideal mirror would be lightweight and yet rigid enough to avoid flexing as the telescope turns. It would be made of a material that could hold a precise parabolic shape, smoothed to avoid unwanted scattered light. The material would also have to be insensitive to changes in air temperature so that its shape wouldn't change from day to night.

Corning Glass Works had fulfilled many of these requirements in 1978 when it fabricated the 2.4-meter blank for the Hubble Space Telescope. Corning built it as a "sandwich" of Ultra Low Expansion glass, in which a thin faceplate and back plate were fused to an "egg crate" of square columns (see fig. 5.1). That design reduced the mirror's weight by 80 percent compared with a solid mirror while preserving its rigidity. In addition, Corning had "slumped" the assembly to create a concave front surface. That greatly reduced the amount of grinding required to reach a parabolic shape.

Angel wondered whether these principles could be carried much further, to mirrors 5 meters and larger. Building a mirror out of separate pieces, as Corning had done, seemed cumbersome to him. There had to be a way to cast a complete blank, with a hollow core, all in one pouring. Angel puzzled over the design of a special mold that would do the job. Eventually he found one he liked. But it lacked one important feature: a way to make a concave faceplate, as Corning had done for the HST. Unless he could find a way, the blank would require a risky grinding operation to make a finished mirror.

After some more thought, Angel found the answer he was seeking: *spin the mold* while the glass was still molten.

Around 1680 Sir Isaac Newton had learned that the surface of a spinning liquid takes the shape of a paraboloid, the ideal shape for a telescope mirror. He actually proposed to build a telescope with a liquid mirror but lacked the technology. Robert Wood, an optical physicist, picked up the idea in 1909. He succeeded in building a liquid mercury mirror half a meter in diameter that yielded reasonably sharp star trails (note 7.2). Now Angel planned to apply Newton's discovery to mirror making.

He would build a mold in a furnace, pour in glass while the furnace was rotating, and allow the glass to cool slowly into a parabolic shape. To support such a thin mirror, he needed a rigid glass core. Ideally this structure should be light and hollow, to allow air to circulate behind the faceplate (note 7.3). Angel decided to use a honeycomb of hexagonal glass columns, fused together into a rigid grid. But what kind of glass, and how do you fuse separate columns reliably?

Angel decided to experiment. Around 1980 he built a small electric kiln in his backyard and heated up a cluster of ovenproof custard dishes. With careful control of the kiln temperature he could get the dishes to fuse into a flat honeycomb structure that had impressive strength. Such a structure could meet his requirements for rigidity, low weight, and open structure.

With this mild success he was ready to try for a larger cluster. He tried different types of glass but zeroed in on borosilicate glass, the basis of Pyrex. Eventually he approached Peter Strittmatter, the director of the university's Steward Observatory, for permission to build a larger kiln on a loading dock. As Nick Woolf recalled, Strittmatter asked why, and Angel told him he wanted to make mirrors.

"How big?"

"Oh, about 20 feet across."

Strittmatter was stunned but went along with the wild idea.

Despite his outward confidence, Angel was feeling his way. He proceeded slowly and systematically. Over a period of two years he gained experience in welding a thin faceplate to a rigid honeycomb of glass cylinders. He had his failures but gradually learned to cast larger and larger mirrors. As early as 1982 Angel and his colleague John M. Hill were confident enough in the process to foresee casting mirrors as large as 8 meters. Their predictions were welcome, but they had to show what they could do.

So in 1983 Angel and his team cast a series of small mirrors in a stationary oven, ending with a 1.8-meter mirror for the University of Calgary. Later that year another 1.8-meter popped out of the oven, destined for the National Optical Astronomy Observatory in Tucson. These mirrors still had relatively thick faceplates.

Then in 1985 Angel built his first rotating oven in a vacant Jewish temple near the university campus. After several trials with this device he was able to cast a 1.8-meter mirror, thinner than ever before, for the Vatican Observatory.

BEATING A PATH TO HIS DOORSTEP

Angel was ready to go for the gold. He began looking for some place on the campus to set up a really big laboratory. Through the local grapevine he learned about a huge vacant space under the east wing of the university's football stadium—an odd place to develop high technology, but convenient and available. With funds from the U.S. Air Force, the National Science Foundation, and the university, he built his lab in 1985.

Angel proceeded to cast larger and larger mirrors with his rotating ovens. The first was a baby 1.2-meter, cast in November 1987 for the Smithsonian Institution. Then in April 1988 the first 3.5-meter mirror was cast, this one for the Astrophysical Research Consortium, a group of six universities that had established a new observatory at Apache Point, New Mexico. This mirror had one of the fastest focal ratios ever achieved for its size, f/1.75, which produces a large field of view (the size of the full Moon) at prime focus. Only eight months later the second 3.5-meter mirror was cast for WIYN, a partnership of the University of Washington, Indiana University, Yale University, and the National Optical Astronomy Observatories. This group chose to place their new telescope on Kitt Peak, Arizona, the site of the 4-meter Mayall Telescope. Then in June 1989 the third 3.5-meter was cast, for the U.S. Air Force Phillips Lab, in Albuquerque, New Mexico, which intended it for tracking satellites.

The news was out: Angel could produce mirrors comparable in size to the largest then in service, at a fraction of the cost, in a fraction of the time, and with focal ratios as short as f/1.75. In fact, simply by changing the speed of rotation of the oven, Angel could adjust the final focal ratio of a mirror to the taste of the client.

Spin-casting molten glass in a complex mold was a major advance in mirror technology. It also sparked a revolution in the way astronomers could satisfy their needs for observing time on a large telescope. In the past, American astronomers who were not on the faculty of the University of California or Caltech relied on the National Observatories at Kitt Peak and Cerro Tololo in Chile for time on big telescopes. The biggest of these had 4-meter mirrors—not as large as Palomar's, but outstanding for their time. The problem was access to these telescopes. As dark-sky astronomy boomed in the 1980s, more astronomers wanted more time for fainter objects. The national 4-meters were oversubscribed by factors of four and five. You were lucky to get a couple of nights in a year.

With the advent of larger, cheaper mirrors, universities could band together, obtain funds from private donors or the National Science Foundation, and build their own observatory. Every man a king!

ANOTHER INSPIRATION

Angel was now ready to tackle really big mirrors, 6 and 8 meters in diameter. But first he had to work out a scheme for polishing such monsters. The problem was the *lap,* the abrasive tool that removes unwanted glass in order to produce a perfect mirror shape. Conventional laps were slightly curved and rigid. No matter how they were designed, they could not match the desired shape of an aspheric mirror everywhere, because the mirror's curvature changes gradually from the center to the edges. As a result, a rigid lap introduces small errors in the mirror's shape, which, if not corrected, produce imperfections in a final image.

How do you get around this problem? One way is to do the best you can with a rigid lap and then touch up the mirror by hand. That's slow, labor-intensive, and therefore expensive. Another way is to work with small laps and change them often so that they can fit the curvature of different parts of the mirror. But it would take forever to polish a 6- or 8-meter mirror using small laps. Worse still, a small lap can introduce long flat ripples in the surface.

If Angel was to grind and polish large mirrors, he had to find a better design for a lap. The solution he and his team found was a lap whose shape *changes* under computer control as it moves across the surface of a mirror blank. They called this a *stressed lap,* and it took them several years to perfect.

The basic idea is to mount the polishing face of the lap on an aluminum baseplate that has aluminum tubes fastened around its edges. Steel bands are tied in a triangular array across the tops of these tubes, and the bands can be tightened or loosened with a system of electromechanical actuators. When the band tension is changed, the plate twists and bends into a precalculated warped shape. The lap's curvature varies under computer control according to its position on the mirror face. In this way a lap with a large area always has the optimum shape for polishing a particular area of the mirror surface. In practice a lap can have a diameter as large as 60 cm. Figure 7.2 shows the stressed lap in action. A stressed lap was used to polish the first three 3.5-meter mirrors mentioned earlier.

Fig. 7.2. The stressed lap in action on the 6.5-meter mirror for the Multi-Mirror Telescope

THE NEXT STEP: A SIX-METER

By 1991 a line of customers was forming outside Angel's office. Several university consortia and private institutions were clamoring for big mirrors—*really* big mirrors. Angel was fully confident that he could start casting mirrors as large as 8 meters in diameter, but first he needed more space and a larger oven. He extended the Mirror Lab under the football stadium to an area of 2,400 square meters and installed a huge new furnace, a polishing facility, and a test setup.

The central feature was the giant rotating furnace dome (fig. 7.3). This mushroom-shaped monster is 7 meters tall and 12 meters in diameter. Its 28-ton upper half is removable, while the lower half is a rotating hearth lined with ceramic tile. The speed of rotation can be adjusted to generate a paraboloid of a chosen focal ratio. Typical speeds are 6 or 7 rpm. The furnace can be heated to over 1,000°C by 2 MW of electrical power.

The first mirror 6.5 meters in diameter was cast in April 1992 for the conversion of the Multi-Mirror Telescope on Mount Hopkins. Remember that the telescope originally consisted of six 1.8-meter mirrors mounted as a cluster, with a total area equivalent to that of a 4.4-meter. Now the University of Arizona and the Smithsonian were about to benefit from Angel's pioneering. They would receive a single 6.5-meter with a very fast focal ratio (f/1.25) to replace their cluster of mirrors. The new mirror would have twice the collecting area and fifteen times the field of view of the old cluster.

Casting began with the assembly of a ceramic mold on the floor of the new furnace. After six months of construction and testing, the mold was ready. Then 10 tons of glass in 2.3-kg pieces were carefully arranged in the mold. The furnace was slowly heated to 1,200°C. In a few days the glass softened and flowed into the ribs of the mold as the furnace rotated at a steady 7.4 rpm. After three months of cooling, the mirror was removed from the furnace, and pieces of the mold were cleaned off. The back face was ground flat and fitted with an array of metal anchors.

As we have seen, a thin faceplate has many advantages over a conventional thick slab. But even with a glass honeycomb to support it, a 6.5-meter faceplate would bend a little as the telescope turns. A sophisticated system was worked out to beat this problem. An array of metal load-spreaders was bonded to the

Fig. 7.3. The spin-casting furnace rotating at 7 rpm

back of the mirror. These objects would mate to a corresponding array of 104 pneumatic actuators in the mirror cell. Under computer control, the actuators would correct any changes in the shape in the mirror due to gravity or wind. Such a system is called *active optics,* an active control of the slow changes in a mirror's shape.

Rough grinding of the front face of the mirror began in September 1995. During 1996 the mirror was polished and tested. The results were very satisfying. The surface of the mirror differed from a precise parabola by no more than 23 nm. To gain some idea of the precision, imagine that we expanded the mirror to the size of the state of Arizona, 600 km across. The errors in the shape of the surface would measure no more than about 2 mm.

Next the team had to lift the finished mirror off the polishing station. This was a ticklish job. How do you pick up a 6.5-meter mirror without bending it unduly or, heaven forbid, damaging its polished surface? The team's solution was an array of suction cups mounted on a heavy steel frame that would attach to the mirror's front surface (fig. 7.4). A vacuum system would exhaust the air from the cups and allow them to bond to the mirror. Then a heavy crane could lift the $10 million chunk of glass from its nest and lower it into its final cell.

The Mirror Lab was also responsible for transporting the precious mirror 90 km south of Tucson to the summit of Mount Hopkins and supervising its

Fig. 7.4. To lift the Multi-Mirror Telescope's 6.5-meter mirror from its polishing station, the crew used suction cups attached to a steel framework.

installation. They would take no chances. They cast a dummy mirror, similar to the finished 6.5-meter, packed it in a special shipping container, and set it on a heavy-duty trailer. The trailer wound its way up the mountain in April 1998. At the top, the dummy was examined for damage. The team breathed a sigh of relief: the dummy had arrived safely. So in July and August the polished mirror was packed up and carried to a base camp at the foot of the mountain. There it rested while the old MMT telescope mount was disassembled and the new hardware was installed.

Finally, in March 1999, the primary mirror was hauled to the summit and installed in the waiting telescope mount. According to Craig Foltz, the director of the MMT, this was no easy task. The new mirror just barely fits into the enclosure. To install it, the 10,000-kg mirror had to be lifted some 20 meters and lowered gently into position. The operation is so ticklish that Foltz hopes never to have to repeat it, even to re-aluminize the mirror. So a special aluminizing rig was designed to work in place, a unique arrangement for telescope mirrors. After the usual shakedown tests and alignments, first light was seen on May 17, 2000. A pair of stars in a binary, separated by only 0.7 arcsecond, were resolved beautifully.

As you can see, casting this huge mirror was only the first step in a long, drawn-out process of converting the telescope. Eight years would pass between casting the mirror and seeing first light through the converted telescope. Several factors accounted for such slow progress. First, Angel and the Mirror Lab team had to feel their way cautiously. At each step they had to invent new methods of grinding and polishing, testing, handling, and transporting their mirror. All kinds of mechanical hardware had to be fabricated to assist in finishing the mirror. On Mount Hopkins, the old mirror mounts had to be replaced and the original enclosure had to be enlarged.

The long struggle was worth it, in the end. The revolutionary spin-casting process had yielded the sponsors of the MMT a mirror at a fraction of the cost of a conventional mirror. The converted MMT (no longer a multi-mirror telescope, but preserving its name) would have a magnificent scientific career.

MAGELLAN

Next in line for a 6.5-meter was the Magellan Project. This is a collaboration among five institutions: the University of Arizona, Harvard, the Massachusetts

Institute of Technology, the University of Michigan, and the Carnegie Institution of Washington.

Andrew Carnegie, one of the great robber barons, established his institution in 1902 with the mission of supporting discovery in the sciences. In 1904 the institution gave George Ellery Hale the money to establish the Mount Wilson Observatory. After Caltech built the 200-inch telescope, with funds from the Rockefeller Foundation, the Carnegie Institution joined Caltech to form the Mount Wilson and Palomar Observatories, which evolved into one of the world's great astronomy centers. In 1969 the institution built an observatory at Las Campanas, Chile, to gain access to the Magellanic Clouds (satellite galaxies to our own), the center of the Milky Way in the constellation Sagittarius, and the rest of the brilliant southern sky. Its principal instrument was a 2.5-meter telescope, built with funds from the Du Pont family.

Around 1980 the Carnegie astronomers began to plan for something bigger. The next step would be costly, however, and the Carnegie Institution was stretched too thin. The institution was supporting a veritable army of scientists in such subjects as embryology, geophysics, terrestrial magnetism, and plant biology, in addition to astronomy. The project languished until news of Angel's success in casting large mirrors came out. That changed the whole landscape.

In 1992 the Carnegie Institution established the Magellan Project, with the University of Arizona and Harvard as partners. They would buy a 6.5-meter mirror from the Mirror Lab, build a telescope at Las Campanas, and use it for the greater glory of astronomy. This was an exciting prospect, but the best was yet to come. Angel realized that he could cast *two* 6.5-meter mirrors for little more than the price of one, once he had the equipment ready. All that was needed was that increment in funding.

New partners were not hard to find. In 1996 the University of Michigan and MIT joined the Magellan Project. Two 6.5-meter telescopes would be built side by side at Las Campanas. One would be named for Walter Baade, the famous Dutch-born astronomer who discovered that stars in our Milky Way divide into two distinct populations (see notes 3.2 and 4.6). The second would be named for Landon T. Clay, a Harvard alumnus and Boston philanthropist. Each partner in the consortium would receive telescope time in proportion to its financial contribution. Carnegie would get 50 percent, the other partners about 10 percent each, and the University of Santiago would receive the remaining 10 percent.

Fig. 7.5. This photo-
graph of the Multi-
Mirror Telescope
casting gives one a
good impression of
the size of the 6.5-
meter mirror for
Magellan I.

With the experience of casting the first 6.5-meter mirror behind them, the
Mirror Lab team felt confident that they could spin out two copies with ease
for the Magellan Project (fig. 7.5). Indeed, they encountered far fewer problems. They cast the Magellan I mirror in January 1994. Grinding and polishing had to wait, however, until the MMT mirror was finished. Not until January 1997 was work on Magellan I resumed. However, the results were worth the
wait. A surface accuracy of 14 nm was achieved, almost twice as good as with
the MMT mirror. Like the MMT mirror, Magellan I has a very fast focal ratio,
f/1.25.

Shipping a 6.5-meter mirror to a mountaintop nearly 10,000 km away turned into a major campaign. The 9.5-ton mirror was packed into a special 14-ton container on November 7, 1999, and loaded on a trailer truck. The mirror cell was loaded onto a second truck and miscellaneous parts onto a third. The convoy of three trucks, accompanied by four pilot trucks and four Arizona highway police cars, crawled toward California. At the border, the California police escort was ready to assist. The parade reached Long Beach, the port of Los Angeles, on November 9. The cargo was loaded aboard a ship on November 12 and arrived at the port in Chile sixteen days later. From there, with many adventures along the way, the mirror was hauled slowly up to the observatory at an elevation of 2,400 meters.

The whole story was repeated with the Landon Clay mirror, which was cast during the autumn of 1998. After three years of grinding, polishing, and testing, the mirror was readied for shipment to Las Campanas. On July 17, 2001, the mirror, nestled in its special 7-meter container, was loaded onto one truck and its mirror cell onto a second truck. Once again the convoy started out for Long Beach, with a police escort to clear the road for the superwide loads. Unfortunately, the direct route to Long Beach from Tucson was blocked by construction, so the convoy had to take a long detour through Yuma, around San Diego, and north up the West Coast. One can only try to imagine the traffic jams such a parade would cause! But with the help of the California Highway Patrol, the mirror arrived safely at Long Beach just past midnight.

Then the mirror was ready for its voyage to Chile. It was loaded aboard a container ship and sailed for fifteen days to the port of Coquimbo. Next, back onto trucks and a 150-km trip to Las Campanas. After a coat of aluminum was applied to the mirror surface in a special facility on the mountain, the mirror was installed in the waiting telescope mount in November 2001.

The Baade Telescope saw first light on September 15, 2000, and the Clay began regular observations on September 7, 2002. Both telescopes are equipped with an array of imaging spectrographs and wide-field cameras, with such imaginative acronyms as MagIC, MIKE, and PANIC.

The Magellan telescopes have an exceptionally wide field of view, 30 arcminutes at the f/11 Nasmyth focus and a 1-degree field at the Cassegrain focus. Eventually the two telescopes will be optically coupled to form an interferometer.

SOME RECENT RESEARCH

Roger Angel takes great pride in his handiwork. The huge new mirrors are in their way a triumph of technology. But as a practicing astronomer, Angel must also delight in the science that some of his glass wonders are turning out.

The Magellan telescopes were among the first of the new generation of giant telescopes, and they are in high demand. They have been used to observe a tremendous variety of objects, including a brilliant supernova, the active centers of galaxies, gravitational lensing of distant quasars, and the dust disks around young stars. As a specific example, consider Type Ia supernovas.

As we learned in chapter 6, two groups are using Type Ia supernovas as standard candles to determine whether the expansion of the universe has changed over time. They both conclude that, rather than slowing down as expected, the expansion is accelerating. The cause of this expansion is thought to be dark energy, a repulsive force that overcomes the attraction of gravity.

This is all well and good, but until recently nobody understood just why Type Ia supernovas explode, or why their maximum light is related to the shape of their light curve. The preferred model for this type of supernova is a binary system consisting of a white dwarf and a massive star. Theorists had suggested that the heavier star might lose mass that spills on to the dwarf, causing it to exceed the Chandrasekhar mass limit (1.4 solar masses) and therefore to explode catastrophically. But nobody had actually observed such behavior until Mario Hamuy (Carnegie Observatories) and Nicholas Suntzeff (Cerro Tololo Inter-American Observatory) caught supernova 2002ic in the act. They obtained the telltale spectrum of a hydrogen accretion disk with the Baade Telescope at Las Campanas. Incidentally, the host galaxy of this key supernova lies at a distance of about a billion light-years.

THE ULTIMATE MIRROR

Angel's next challenge was to cast two mirrors 8.4 meters in diameter for the Large Binocular Telescope (LBT). This is a project of a consortium with many partners, including three universities in Arizona, four in other U.S. states, six in Italy, and five in Germany. The telescope will be located on Mount Graham, a pristine 3,200-meter mountain in southeastern Arizona. Unlike the Magellan telescopes, which have independent mounts and domes, the LBT mirrors

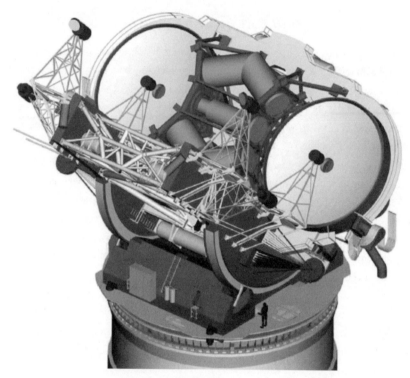

Fig. 7.6. The mount for the Large Binocular Telescope stands 40 meters tall. Note the human figure at the lower right in this rendering.

will be mounted side by side on a single huge alt-azimuth mount in a single enclosure (fig. 7.6). The telescopes will either be used independently or be coupled as an interferometer. Working together, the LBT mirrors will have an area equal to an 11.8-meter.

Peter Strittmatter, director of the Steward Observatory, was the driving force behind this project. The Steward Observatory had built a 90-inch (2.3-meter) telescope atop Kitt Peak in the mid-1970s. A decade later, with Angel's spin casting a proven technique, Strittmatter began to think bigger. Angel had already forecast the possibility of spin-casting mirrors as large as 8 meters in diameter, and that ultimately is what Strittmatter aimed at.

He had a long list of obstacles to overcome. Among the most pressing was the choice of a site. There were many possible sites in the desert Southwest within easy commuting distance of Tucson, and Strittmatter saw no overriding reason to look further. He selected Mount Graham from a list of 280 candidates as having the best combination of weather, dark sky, a dry climate, and good, if not superb, seeing.

The mountain was part of a 62,000-acre wilderness, however, and therefore closed to development. So Strittmatter enlisted Barry Goldwater, senior senator of Arizona, to help him obtain access to the mountain for an observatory. In 1988 an act of Congress authorized a 24-acre tract for an observatory on the mountain, with a limit of seven telescopes. The University of Arizona would have to build a road to the top of the mountain and to provide power, water, and shop facilities.

But Mount Graham is the home of an endangered species, the Mount Graham red squirrel. The Sierra Club and its associates initiated a series of lawsuits and appeals to protect the squirrel and, incidentally, to halt construction in a designated wilderness. Another coalition of environmentalists launched a second round of suits and appeals. After five years of litigation, the project was allowed to proceed by the courts. The San Carlos Apache Indian tribe is still protesting the desecration of the sacred sites on Mount Graham, however.

Strittmatter's second task was to round up partners who could share the cost of building and operating the huge telescope. He was untiring in his efforts and, as we saw, marvelously successful. To reflect the diversity of the partnership, a new observatory name, the Mount Graham International Observatory, was adopted. Strittmatter's Steward Observatory, however, ends up with only 7 percent of the total observing time. The opportunity to observe on this telescope is priceless, but when so many partners are involved, each gets only a small crumb.

HOW TO BAKE A GLASS CAKE

Each step in the casting of the first 8.4-meter mirror had to be worked out meticulously. Every detail had to be nailed down to ensure a nearly perfect final product. And even so, casting an 8.4-meter had its perils. Air bubbles, cracks, and uneven flow of the glass were constant worries.

The casting began with the preparation of a mold on the furnace hearth. (Fig. 7.7; the four panels of the figure illustrate the process for the second 8.4-meter mirror, but the same procedure was followed for the first mirror.) A

Fig. 7.7. Several steps in the casting of the second mirror for the Large Binocular Telescope are shown in this mosaic. *Top left,* laying the tiles on the floor of the hearth. *Top right,* installing the core boxes for the honeycomb. *Bottom left,* laying out 20 tons of glass. *Bottom right,* the furnace spins at 7 rpm, day and night.

floor of refractory tiles was laid down to define the precise shape of the mold (fig. 7.7, top left). Next, a circular wall of forty-eight silicon carbide panels was built around the edge of the mold, and a ceramic cylinder was set up at the center to create a hole through the mirror. This stage of assembling the mold required six months of labor.

After a preliminary bake-out of the mold, strong bands of a high-temperature nickel alloy were wrapped around the outside perimeter of the panels. These bands would resist the outward pressure of the molten glass and still allow the mold to expand and contract as the heating and cooling cycle progressed. At this stage of the construction, the entire mold was heated to join the bands and the walls.

Then a special refractory layer was glued to the mold's interior and machined to an exact shape. The mold was now ready to receive the hollow *core boxes* that form the glass honeycomb behind the faceplate. These core boxes had a hexagonal cross-section and varied in length from about 1 meter at the outside edge of the mold to 50 cm near the center. They were made of a silica-alumina fiber that can withstand both the heat and the pressure of the casting process, and still crumble easily after the casting is finished. The 8.4-meter mirror required almost two thousand cores, each bolted a precise location to the floor of the hearth (fig. 7.7, top right).

At last the mold was ready. Twenty tons of presorted glass chunks were laid out on the tops of the cores (fig. 7.7, bottom left). Angel had chosen a variety of borosilicate glass similar to Pyrex but manufactured specially by Ohara, a Japanese firm. It has many advantages. It expands and contracts very little with a change of temperature, it melts and flows easily at reasonably low temperatures, it is fairly cheap, and it resists the acids that would be needed to periodically remove the aluminum coating of the final mirror.

In January 1997 the top of the furnace was sealed. The computer took over, raising the temperature slowly and maintaining a constant speed of rotation of 6.8 rpm (fig. 7.7, bottom right). Over a period of five days the temperature rose to 1,180°C. The glass melted, and part of it flowed down in the spaces between the core boxes to form the ribs. The remainder of the molten glass collected in a pool over the tops of the cores to form the parabolic surface of the mirror. When the glass had slowly cooled to 600°C, the rate of cooling was slowed even further to anneal the glass. (*Annealing* relaxes the stresses that build up as the mirror cools unevenly.)

In April 1997 the mirror team had its first look at the mirror. There was a problem. A few pieces of the mold's wall must have shifted, allowing glass to drain from the top pool. The faceplate of the mirror would be too thin unless some remedy was found. The team decided to add 2 tons of glass to the furnace and raise the temperature rapidly. In this way the additional glass would melt and flow before the main body of the mirror softened. In September, after three months of cooling, the team had another look, and this time they were satisfied. The faceplate had a uniform thickness of 4 cm.

The mirror was allowed to cool down to room temperature, with the lid of the furnace slowly rising. Next, it had to be lifted out of the furnace and moved to a cleaning station where high-pressure jets of water would remove the cores of the mold. Handling a mirror this size is no small matter. It weighs over 20 tons and is as fragile as a wineglass. Smaller mirrors had been lifted by attaching suction cups to the surface and hoisting them with a heavy-duty crane. The team worried that this scheme might not work with the 8.4-meter. So an ar-

Fig. 7.8. The first 8.4-meter mirror blank is ready for grinding. John Hill, casting scientist, stands at lower left.

ray of pads were glued to the surface with special cement and attached to a steel framework that the crane could grab. There were problems with the glue, but in the end, everything worked.

Next the mirror went to the grinding station. The mirror was turned over, face down, and had its backside ground flat. That took six months, from January to June 1999. Figure 7.8 shows the largest mirror ever cast, a 20-ton jewel more valuable than diamonds. In May 2000 the second 8.4-meter mirror was cast. This time there were no glass leaks, and the mirror came out of the mold in excellent condition.

Work on the first mirror picked up again in the autumn of 2000. The next task was the installation of so-called load-spreaders, which are triangular metal fixtures that distribute the weight of the mirror in its final cell. The spreaders were bonded to the back of the mirror with RTV, a powerful adhesive, during the winter of 2000.

Then in March 2001 the mirror was flipped over and set in a special cell for grinding. That lasted between August and October 2001. At last the mirror was ready for its final polish, with a stressed lap. By September 2003 the second mirror had been polished and was being lifted into its cell.

PROGRESS TOWARD THE FINISH LINE

The primary mirrors are perhaps the most critical parts of the Large Binocular Telescope, but building the telescope mount was also a challenge. The alt-azimuth mount will stand 40 meters high above bedrock and weigh 580 metric tons. The two primary mirrors swing in *elevation* (angle above the horizon) in two large C-shaped arms, which turn on a horizontal ring (see fig. 7.6). Above the mirrors stands a large, somewhat awkward-looking structure that allows secondary mirrors to rotate into position to change the optical configuration.

The design, prefabrication, assembly, and testing of the huge structure was divided among several competent Italian firms. They completed the telescope mount on schedule and shipped it to Arizona in July 2002. The bearings and gears followed later.

On Mount Graham, an enclosure for the telescope was completed in early 2000. The very fast focal ratio of the primary mirrors (f/1.142)—or, equivalently, the very short focal length (9.6 meters)—led to a compact, economical design. Nevertheless, the building is huge, 40 meters high.

By October 2003 most of the mechanical and electrical components of the Large Binocular Telescope had been installed in the enclosure. In March 2004 the first mirror and its cell were installed on the telescope (a Herculean task), and the first stellar images were obtained in September. The second mirror was still being polished as of June 2004.

The Large Binocular Telescope is clearly on the path to completion. On October 25, 2004, the LBT was dedicated and a formal dinner was held to mark the occasion. A large crowd was invited to tour the newly completed telescope. First light with both mirrors is eagerly awaited. There will follow a normal period of testing, fine-tuning, and debugging.

The LBT should begin scientific observations sometime in 2005, barring

any unforeseen problems. The world will gain a unique instrument for probing the universe.

Roger Angel has moved on. He has spent two decades developing the art of
spin casting and overseeing the casting of huge mirrors. But an innovative
mind like his cannot be constrained indefinitely to one project, however large.
He recently established the Center for Astronomical Adaptive Optics at the
University of Arizona. (We'll have more to say about adaptive optics in a later
chapter.) He is planning how to use the Large Binocular Telescope as an interferometer. He is also thinking about the next generation of giant telescopes,
30 meters and larger, both on Earth and in space. With another team he is exploring new ideas for electrostatically shaped membranes for space telescope
mirrors. He is deeply involved in designing the 8.4-meter Large Synoptic Survey Telescope, which would be used to map the entire visible sky digitally. And
occasionally he spins out an idea for a novel spectrograph.

Lord knows when the man sleeps.

A PROFUSION OF TELESCOPES

In the 1980s, telescopes with a variety of new features captured the attention of astronomers around the world. Among them were the Multi-Mirror Telescope in Arizona, the New Technology Telescope in Chile, and the Advanced Technology Telescope in Australia. These introduced such innovations as alt-azimuth mounts, thinner mirrors, active optics, and special enclosures.

At the same time, a revolution in the art of making mirrors was under way. Jerry Nelson was building the 10-meter segmented mirror for the first Keck telescope, and Roger Angel was about to spin out 6.5-meter monolithic mirrors. At such industrial firms as Corning Glass in the United States and Schott Glaswerke in Germany, optical engineers were experimenting with other daring techniques.

In the 1990s these two streams of invention came together. Now it was possible to build an advanced-technology telescope, with all its desirable features and with a mirror 8 meters or larger. Between 1985 and 1995, eight giant telescopes were completed, and construction on three more began. White domes were popping up like mushrooms after a rain.

In a way this stampede toward huge mirrors was surprising. Photon detectors and digital cameras were more sensitive than ever, especially in the infrared, and were making more efficient use of the available light. Similarly, fiberoptical spectrographs and imaging spectrometers were allowing astronomers to carry out un-dreamed-of research programs on "mere" 4-meter telescopes.

So what drove this stampede toward massive light-gathering power? The most obvious reason is that astronomers never have enough light, no matter what they seek. Some astronomers wanted to push as deep into space as possible, to view the universe as close to its moment of creation as possible. Cosmology was booming, and to compete, a nation needed the largest mirror it could afford, in a state-of-the-art telescope, at a world-class site. At the same time, the hunt for extrasolar planets was finally bearing fruit, with exciting prospects for finding life somewhere in the vast spaces. Infrared astronomy was also blossoming. Astronomers would now be able to penetrate the dense interstellar clouds surrounding infant stars, to learn how stars are created. For such research, more light meant more science.

Not only could the new telescopes be larger, they could also be much more capable. Astronomers would soon be coupling the new telescopes, in pairs or groups, to form optical interferometers. That would bring on another revolution, in angular resolution. Structures a hundred times smaller than ever before could be studied within a distant galaxy.

For all these reasons, astronomers around the world pressed for a massive program of telescope construction. In the decade from 1985 to 1995, their dreams were fulfilled.

THE VERY LARGE TELESCOPE

The European Southern Observatory (ESO) was established in 1962 as a joint enterprise of eight nations, which eventually grew to ten. As we saw in chapter 4, this organization founded the La Silla Observatory in the arid foothills of Chile. The seeing was excellent, the nights dark and cloudless. La Silla was gradually equipped with several telescopes of moderate size and then two in the 4-meter class. Over two decades the observatory earned a fine reputation for scientific productivity, and its sponsors were well pleased.

By 1980, ESO astronomers recognized that they would need a larger telescope if they were to fully exploit the potential of the Chilean Andes. They decided to proceed in stages. They first built the 3.6-meter New Technology Telescope at La Silla to gain experience with all the new hardware. The NTT was completed in 1989 and met all the designers' expectations. They had mastered the difficult technology of the new age and were ready to move on to the next stage, a giant telescope. The first questions, as always, were how large a mirror to aim for, and who could supply it.

Roger Angel was confident that he could cast 6- and 8-meter giants, but he hadn't actually demonstrated that he could. In fact, he would cast his first 6.5-meter only in 1992. ESO astronomers were not willing to wait several years.

The construction of the first Keck telescope was well advanced in 1989, but the quality of its images would be untested until 1992, when all thirty-six segments were installed. The astronomers of ESO were intrigued by Nelson's design, but understandably cautious. They preferred a more conventional approach. They contacted the Schott Glaswerke, in Mainz, Germany, a subsidiary of the world-famous optics firm Carl Zeiss, and asked what the company could do for them.

According to Hartmut Hoeness, leader of the Schott team (*Astrophysics and Space Science* 160, 193, 1989), the company had been spin-casting mirrors of its low-expansion ceramic Zerodur (note 8.1) for several years. The team began by casting a 1.8-meter blank and, as they gained experience, cast blanks as large as 4.1 meters. Unlike Angel's Mirror Lab, Schott was spin-casting monolithic mirrors, without a hollow core, in the shape of a thin meniscus. The Schott team was confident that they could cast mirrors as large as 8 meters, but they would need time and money to expand their facilities.

This alternative was still a gamble, but it looked like the best available to the ESO astronomers. They reached a consensus and drafted an ambitious conceptual design based on Schott's estimates. They submitted their proposal to the governing council of ESO, and after due deliberation the council responded.

In December 1987 the council voted to build the largest telescope in the world. It would consist of four 8-meter telescopes mounted side by side in separate enclosures. They could be used separately or combined as an interferometer. In this interferometer mode, the Very Large Telescope (VLT) would have the light-gathering power of a 16-meter.

Each of the four units would have Ritchey-Chrétien optics, with an f/13.4 Cassegrain focus, two f/15 Nasmyth foci, and a Coudé focus (note 8.2). Each telescope would be equipped with active optics to correct the aberrations of its mirror. The telescopes would be housed in enclosures carefully designed to preserve the good seeing of the site. In addition, ESO would build up a suite of advanced instruments to utilize the great power of the telescopes.

ESO was prepared to spend at least $250 million to obtain the finest observatory on the planet. Most of this money would be spent within the member nations, an arrangement that helped persuade them to participate. The council set a target date of 1999 for completion of this enormous project.

Early in their planning the ESO astronomers had anticipated that the summit of La Silla might be too small for a large new telescope, so they began to search for alternate sites. (In chapter 11 we'll learn about the refined techniques used to test the seeing at the sites.) After several years of testing, they narrowed their choices to two sites. The first was a mountain conveniently near La Silla. The second was a 2,600-meter mountain, Cerro Paranal, some 120 km south of Antofagasta. Cerro Paranal proved to have somewhat better characteristics and was finally selected for the project.

In 1988 ESO signed a contract with Schott, and the company built a rotating furnace for mirrors 8 meters and larger. Production of the mirrors began in 1993. Schott's procedure was more complicated than Roger Angel's. It required risky transfers of the mirror blank between two furnaces, as well as a long annealing process.

The Schott engineers began by melting 45 tons of Zerodur in a concave mold inside a stationary furnace (see note 8.1 about Zerodur). The molten load was then transferred to a rotating furnace. As the furnace spun, the temperature was slowly reduced until the Zerodur solidified. Then the meniscus blank was transferred to an annealing oven to relieve the stresses built up as the blank cooled. Each 8-meter blank needed three months for this process.

After a blank cooled to room temperature it was removed with suction cups from the mold and turned over to expose its convex surface. A crystalline layer, which has a different coefficient of expansion than the rest of the blank, had to be ground off first. Then the blank was flipped over, and the concave surface was ground. Next, the blank was returned to the annealing furnace and cooked for nine months. This step was necessary to convert the internal structure of Zerodur from its initial glassy state to its low-expansion ceramic state.

The pace of production was impressive. The first blank was cast in July 1993, almost four years before Angel cast his first 8.4-meter for the Large Binocular Telescope. The second casting came out of the furnace in November 1994, the third in September 1995, and the fourth in September 1996. Each blank was 8.2 meters in diameter and 17.5 cm thick and weighed 23 tons. Schott delivered the blanks to the French firm REOSC Optique for final grinding and polishing.

As mentioned earlier, the cells that hold the mirrors were provided with actuators to eliminate the mirrors' aberrations. REOSC had to install 150 connections on the backside of each mirror for these actuators.

Fig. 8.1. Paranal Observatory as of 1999. The tracks under construction in the foreground are part of the Very Large Telescope interferometer.

Meanwhile, other contractors were busy. One group began preparing the site in 1992. By 1993 some 300,000 cubic meters of rock had been removed from the top of Cerro Paranal to provide space for the four giant enclosures (fig. 8.1). The Swedish firm Skanska-Belfry Ltd. was employed to build the telescope foundations and buildings. Ansaldo Energia, an engineering firm in Milan, was responsible for fabrication of the 400-ton telescope mounts (fig. 8.2).

In the midst of all this frantic activity, ESO was hit with a lawsuit. The Chilean government had donated 725 sq km of land surrounding Paranal Mountain to ESO in 1988, with the understanding that the VLT would be built there and that Chilean astronomers would be able to use the telescopes. In March 1993, however, the heirs of Admiral Juan José Latorre claimed that Paranal was part of their inheritance. The admiral had been given the land in gratitude for his services in the wars with Peru in the late 1800s, and the heirs wanted it returned.

The case provoked a public outcry against the foreigners, and the council of ESO was embarrassed. ESO invoked its immunity against prosecution under the terms of a treaty with the Chilean government at the time that La Silla

Fig. 8.2. One of the 8.2-meter telescopes on Paranal

was procured. The case was simply an internal Chilean matter, the council claimed.

The suit moved up to the Chilean Supreme Court, which decided against the Latorre family. Following a meeting of the Chilean government with ambassadors from ESO's collaborating countries, a supplementary treaty was signed, which allotted more observing time to Chilean astronomers and set rules governing the number of telescopes to be built at Paranal. The VLT was back on track again.

Construction of the four enclosures on Paranal was completed early in 1997. The first mount was installed during the last half of 1997 and the first mirror in 1998. On May 27, 1998, the first 8-meter reached an important milestone: first light. A number of familiar objects were photographed as a test of the whole system. The astronomers were delighted with the results. The long, complicated, expensive effort had paid off.

In March 1999 the completed Paranal Observatory was dedicated at a ceremony on the mountain. The four telescopes, previously known as Units 1, 2, 3, and 4, were given the names of objects in the sky, in the Mapuche language of a native Chilean people. The names are now Antu (the Sun), Kueyen (the Moon), Melipan (the Southern Cross), and Yepun (Venus).

Each telescope has been equipped with a different instrument for each of its three focal planes. Antu, for example, has CRIRES, a high-resolution infrared spectrometer at Nasmyth A; FORS1, a low-resolution spectrograph at the Cassegrain; and ISAAC, an infrared spectrometer and array camera, at Nasmyth B.

Since 1999 the four telescopes of the VLT have been used to churn out an impressive stream of cutting-edge science. In October 2002, for example, a team from the Max Planck Institute for Extraterrestrial Physics announced that they had obtained the orbit of a star that circles the supermassive black hole at the center of our galaxy (fig. 8.3). The star, called simply S2, approaches the hole to within a distance of only 17 light-hours, or 130 times the distance from the Earth to the Sun. S2 orbits the hole at a speed of 5,000 km/s and makes a complete circuit in only fifteen years. This observation helps to confirm what had been suspected for some time: that the massive object at the galaxy's center is not a cluster of unusual stars or exotic particles but rather a black hole with a mass greater than that of two million Suns.

THE JAPANESE GO ELSEWHERE

In 1980 the largest telescope in Japan was the 1.88-meter at Okayama Astrophysical Observatory. Japanese optical astronomers had been waiting to build a 4-meter since 1974, but their national research organization had decided to build a world-class radio telescope at Nobeyama first. That was probably a

1 light-year

Fig. 8.3. A highly resolved infrared view of the center of the Milky Way, obtained in 2002 with Yepun, one of the units of the Very Large Telescope. The length of the bar indicates a distance of 1 light-year. We can see stars of different ages, some still enveloped in their clouds of dust. The two arrows point to Sagittarius A, the exact center of the Milky Way, where a supermassive black hole resides.

sound decision, because Japan's climate is not well suited for optical astronomy. The nighttime astronomers persisted, however, and in 1984 they got a green light. Their proposal to build a telescope larger than 5 meters, in Hawaii, was accepted. They would design an instrument optimized for the infrared, with all the modern features of active mirror control and advanced enclosures.

The National Observatory of Japan, representing many Japanese universities, was organized in 1987 to oversee the project. Keiichi Kodaira, an expert in galactic physics, was appointed its director. Mauna Kea was his first choice for a site for the observatory, so he began negotiations immediately with the University of Hawaii, which supervises access to the mountain. In time the Japanese were awarded an area near the summit.

The most critical issue, and the most contentious, was the size and type of mirror. The experts considered several designs, including honeycomb, egg-crate, thin meniscus, and bored-out structures. After much debate they chose the most conservative approach, a thin meniscus, at least 7.5 meters in diameter, with a focal ratio of f/2.3. The telescope would have Ritchey-Chrétien optics, with an additional corrector lens at prime focus to provide a field of view of 0.5 degree.

Schott Glaswerke was a possible source for a spin-cast meniscus, but the firm was fully occupied in casting the mirrors for the Very Large Telescope. The Japanese would have to wait their turn. At this point the Corning Glass Company entered the picture.

Corning had cast the 200-inch (5-meter) mirror for the Palomar Observatory back in the 1940s and had fabricated the mirror blank for the Hubble Space Telescope. Since then the firm had acquired extensive experience in fabricating telescope mirrors of moderate size, using Ultra-Low Expansion (ULE) glass, a patented compound of theirs. They had also developed the technique of heating a flat circular blank and allowing it to slump into a segment of a sphere. In this way they could approach the desired parabolic shape without a great deal of grinding, and with great savings in time and money. This was a well-tested procedure for moderate-sized mirrors. Corning offered it to the Japanese as an alternative to spin casting.

The Japanese were interested. Corning promised that it could provide a mirror at least 8 meters in diameter, with a focal ratio as small as f/2. This was an exciting goal, so the Japanese accepted and the project was under way. In the end it took Corning three years (1991 to 1993) to tool up for the job.

In the summer of 1994 Corning successfully cast an 8.3-meter blank 20 cm thick. It was the largest monolithic mirror blank in the world, predating Angel's 8.4-meter for the Large Binocular Telescope by thirty months. The process began by laying out fifty-five pieces of ULE in a mosaic and fusing them into a single blank. The blank was then slumped into the desired curvature. After an extended period of cooling to room temperature, the blank was shipped to Contraves Brashear Systems in Pittsburgh, Pennsylvania, for grinding and polishing.

Unlike Angel's honeycomb design, the Corning blank had no provision for mounting the actuators that would maintain the shape of the flexible mirror. So 261 holes had to be drilled into the back of the blank—a delicate operation. Four years were required to complete the mirror, but the results were worth the wait. The mirror surface is accurate to a mere 12 nm, or about fifty times smaller than the width of a human hair.

With the production of the mirrors under way, Japanese engineers could proceed with the design of the telescope mount (fig. 8.4). It would be a standard alt-azimuth, with the usual four foci: prime focus, Cassegrain focus, and two Nasmyths. The finished mount would stand 22 meters high and weigh 500 tons.

The Japanese were meticulous in choosing a site for the telescope. They built a scale model of the summit of Mauna Kea and tested it in a wind tunnel. Not surprisingly, they determined that a location near the Keck telescopes

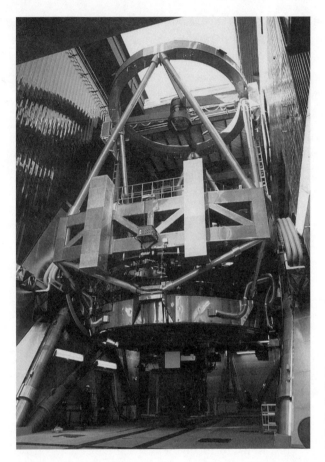

Fig. 8.4. The Subaru telescope has a standard alt-azimuth mount with an 8-meter meniscus mirror.

would be nearly ideal. The next question was how high to build the enclosure. They set up a 30-meter tower, equipped with sensitive thermometers at intervals along its length, to determine the fluctuations of temperature as a function of height. Such measurements indicate the amount of turbulence above the ground. After four months of testing they decided that a height of 27 meters would be adequate.

The telescope enclosure also received special attention. The engineers knew, from long, painful experience, that a flow of warm air within an enclosure could ruin the seeing. So they carried out computer simulations and experiments in a wind tunnel with various small models to determine the best shape for a smooth circulation of air. The result was a cylindrical structure, air-conditioned by day and provided with an array of fans to circulate air at night.

Assembly of the telescope mount within the enclosure began in October 1996 and was completed in March 1998. The precious 23-ton mirror was hauled to the summit and installed in October of that year. In order to reach the top of the mountain, the winding "saddle" road from Hilo, the largest city on the Big Island of Hawaii, had to be widened. In several places road signs had to be removed. The dangerous trip took three days.

In parallel with all this activity on the mountain, the Japanese built a base in Hilo from which to operate and maintain their national telescope. The Hilo facility provides office space for several dozen staff, as well as shops and laboratories in which new instruments are being built and tested. Several supercomputers were also installed to process the stream of observations that the new telescope would produce.

At last, in January 1999, the new telescope saw first light. It was given the name Subaru (no connection to the automobile company), the Japanese word for the Pleiades, the famous constellation of seven naked-eye stars. The test images were spectacular and justified the long effort. The Japanese had their national telescope. They installed the first of a battery of sophisticated infrared imagers and settled down to observe the cosmos.

In May 1999, just four months after first light, Subaru made its first discovery. A team of astronomers made infrared images of a star-forming region, S106, a mere 2,000 light-years from Earth. At the center of the region lies a star only one hundred thousand years old, with a mass of twenty of our Suns. The star, called IRS4, is ejecting hydrogen gas at a prodigious rate as it continues to collapse (fig. 8.5). Surrounding the star is a rotating disk of gas that may, in time, break up into a collection of planets. Many objects like this had been seen before, but rarely in such detail.

Hundreds of faint young objects surround this massive central star. The largest has a mass only one-tenth that of the Sun, about the mass of Jupiter, and is too small ever to become a self-luminous star. These objects, *brown dwarfs*, are not orbiting the star but are being born in its neighborhood. Brown dwarfs may be very common, and some astronomers have proposed that they constitute the nonluminous "missing mass" in galaxies.

Fig. 8.5. A star in the throes of birth. IRS4 is ejecting powerful winds of dust and gas as it contracts. This image is one of several obtained by a team at Subaru in May and June 1999.

A TWIN IN EACH HEMISPHERE

When Jerry Nelson and colleagues published their work on stress polishing and active segment control, astronomers around the world pricked up their ears. Here was an intriguing technology for building mirrors much larger than was ever possible before.

The National Optical Astronomy Observatory (NOAO), based in Tucson, Arizona, decided to gain some experience with the technology. This organization is operated by a consortium of about forty universities and is chartered by the National Science Foundation to build and operate telescopes for American astronomers who lack such facilities. The NOAO was created by merging the Kitt Peak National Observatory (KPNO) near Tucson with the Cerro Tololo Inter-American Observatory in Chile. In the early 1980s each site had a 4-meter telescope in operation, and both were heavily oversubscribed. The managers of NOAO recognized that they had a responsibility to provide more observing opportunities on a first-class telescope. But like the Californians, they learned that Congress, their principal source of funds, would only be interested in funding the largest possible telescope.

The idea of a building a giant telescope had been stirring long before Nelson appeared on the scene. In the mid-1970s Leo Goldberg, the director of KPNO, proposed to build a 15-meter telescope as a cluster of four mirrors, something like the Multi-Mirror Telescope at Mount Hopkins, Arizona. His "Next Generation Telescope" was considered too risky by the advisers to the National Science Foundation, however, and was quietly tabled.

With the passage of time and with improvements in technology, a 15-meter telescope began to look feasible to the director, Geoffrey Burbidge. In the early 1980s, engineers and astronomers at NOAO began to discuss the design of a so-called National New Technology Telescope (NNTT). Two competing designs emerged from the discussions: a cluster of perhaps six 7-meter mirrors on a single mount, or a single 15-meter segmented mirror that would utilize the new technology conceived by Nelson and associates.

As a first step, the staff of KPNO decided to investigate stressed mirror polishing in their optics shop. They would attempt to stress-polish a 2-meter segment. Nelson was eager to cooperate and acted as a consultant for the effort, which could only help in his own work. The experiment was a success, but the prospect of building a 15-meter segmented mirror still looked too daunting.

A telescope similar in concept to the Multi-Mirror Telescope seemed more feasible. Jacques Beckers, a solar astronomer and a talented designer of instruments, led the design team. He had previously worked on the MMT and in fact was responsible for co-phasing its four mirrors to act as a single 16-meter telescope. The great advantage of such a design is that it produces bright, high-resolution images over a wide field of view.

For the NNTT, Beckers envisioned an array of four mirrors, each 7.5 meters in diameter. When assembled on an alt-azimuth mount and co-phased, this quartet could have the angular resolution of a single 21-meter mirror. By 1985 he and his team had developed the internal optics needed for co-phasing and had tested them on a mock-up. Beckers and colleagues continued to develop the design over the next three years and to promote the project for its scientific potential. But NOAO's board of directors were concerned about its complexity and estimated cost. They put a hold on the project. As a result, Beckers left the organization in 1988 to work for ESO and the VLT project.

In 1987 Sidney Wolff was appointed as the new director of the NOAO, and she changed course. She convinced the majority of nighttime astronomers and her board of directors that two 8-meter telescopes would be easier to build than one with four mirrors. One telescope would be located in the Northern Hemisphere, its twin in the Southern Hemisphere. This concept was the birth of the Gemini project, named for the constellation that contains the twin stars Castor and Pollux.

Each 8-meter mirror would be a single monolithic casting, similar to those of the VLT. The telescopes would have Ritchey-Chrétien optics, with a prime focal ratio of f/1.8, and Cassegrain and Nasmyth ratios of f/16 and f/19.6.

From the outset, NOAO planned to optimize the telescopes for infrared observations at wavelengths between 1 and 30 microns. Great Britain had already built the United Kingdom Infra Red Telescope on Mauna Kea. With a mirror 3.8 meters in diameter, it was the largest telescope devoted exclusively to infrared research. But with the new mirror technology, 8-meter mirrors were now feasible. In addition, more sensitive detectors of infrared light were becoming available. Astronomers around the world were eager to look through this new window in the spectrum. The times were ripe for more powerful instruments. These considerations had led the Japanese to build Subaru.

Wolff initiated design studies for the telescopes and soon learned that the pair would cost in the neighborhood of $180 million. This was a staggering sum for those times and would require mobilizing the astronomical community to urge Congress for funding. Patrick Osmer and Fred Gillette, senior astronomers at KPNO, took on the job of preparing a scientific rationale and a technical proposal to the National Science Foundation, the sponsor of NOAO. After due deliberation, the NSF announced that it would grant half of the request on the condition that other partners would contribute the remainder. Wolff immediately opened negotiations with several foreign astronomical associations.

In the end, Argentina, Australia, Brazil, Canada, Chile, and the United Kingdom agreed to join the United States in funding, building, and operating these two telescopes. The partners would share the cost not only of the telescopes but of the spectrographs and cameras that astronomers would need to pursue their research. Much of this construction would be carried out in the home countries.

An international organization, the Gemini Observatory, was established to oversee the project. Not surprisingly, Mauna Kea was preferred as the location for the Northern Hemisphere telescope. After the usual lengthy negotiations, NOAO was awarded a site near the summit by the University of Hawaii, which would take 10 percent of the telescope's observing time as compensation. A groundbreaking ceremony was held at the summit in October 1994.

Cerro Tololo could have been an ideal site for Gemini South, the Southern Hemisphere telescope. It was known to have excellent seeing and an arid climate, ideal for infrared observations. Most important from the financial point of view, all the infrastructure, the roads, power, and housing were already in place. Unfortunately, there wasn't enough space for an 8-meter telescope. So a neighboring mountain, Cerro Pachon, was chosen instead. At an elevation of 2,700 meters above the Atacama Desert, this location would do very well, despite the cost of equipping it.

The Gemini consortium had a choice of several possible suppliers of the two mirrors. The Stewart Observatory's Mirror Lab was still spinning out 8-meter mirrors for the LBT, so Gemini would have to wait. On the other hand, Schott Glaswerke, the firm preferred by the European Southern Observatory, had spin-cast four 8-meter mirrors for the Very Large Telescope, and were ready to take on more work. Corning Glass had successfully cast an 8-meter for the National Astronomical Observatory of Japan, and it too was open for business. In the end, the Gemini project chose Corning, as offering the best tradeoffs among risk, time, and money.

In the summer of 1995, Corning engineers laid out forty-two hexagonal blocks of ULE, the low-expansion glass, in its 8-meter furnace. The first mirror blank popped out of the oven in October, a nearly perfect slab 8.1 meters in diameter and 30 cm thick (fig. 8.6). The second blank was fused in January 1996, only three months after the first.

The first blank was rough-ground on all surfaces, then reheated and slumped to form a segment of a sphere. Next, it was ground to a more precise shape with diamond abrasives. The 25-ton blank was then boxed and shipped to REOSC Optique, in Paris, for final polishing to a hyperboloid with a focal ratio of f/1.8.

Fig. 8.6. A technician in a white suit inspects the 8.2-meter mirror for Gemini North at the Corning Glass Works.

REOSC finished the first mirror in February 1998 and shipped it through the Panama Canal to Hilo, Hawaii. From there it was trucked to the summit of Mauna Kea in three days in June. Meanwhile, the telescope mount and the mirror cell were installed in an enclosure designed and built by the Canadians.

Telescope mirrors are coated with a layer of aluminum a few microns thick to make them highly reflective to visible light. The Gemini telescopes, however, are intended to focus infrared light, and for this purpose a silver coating is a better choice. Unlike aluminum, silver tarnishes, especially in sea air. So despite a protective coating, the silver surface will have to be refurbished periodically. For this purpose, the Royal Observatories of the United Kingdom, a member of Gemini, designed a special coating facility capable of handling the delicate 8-meter mirror. This unique tool was installed in the basement of the enclosure atop Mauna Kea, no mean feat.

Like all thin mirrors, the Gemini mirror requires an active support mechanism to keep it from flexing as the telescope turns. The British members of the consortium took on the job of fabricating the mirror cell, with all its actuators, and shipped it to Hilo.

Gemini North, as it was called (fig. 8.7), saw first light in November 1998, a little over three years from the time Corning had turned on its giant oven. That is warp speed in this business. In November 2002, Gemini North was renamed the Frederick C. Gillette Telescope, to honor one of the pioneers of infrared astronomy and a longtime staff member of NOAO. Gemini South, the telescope on Cerro Pachon, advanced just as quickly. Its polished mirror arrived in Chile in March 2000 and was dedicated in December 2001.

As an example of the science that is being done at the Gemini telescopes, I've picked a recent observation by an international team led by Daniel Stern, from the Jet Propulsion Laboratory. They used a cryogenically cooled multi-

Fig. 8.7. Gemini North (now the Frederick C. Gillette Telescope) in its dome at Mauna Kea

object spectrograph at Gemini South to obtain spectra of a quasar with a wavelength shift of $z = 5.77$. This object is the most distant radio source known and among the most distant x-ray sources. Because of its high redshift and the obscuration of the nearby gases, this quasar is invisible at optical wavelengths.

The big surprise was the discovery of spectral lines of carbon, oxygen, nitrogen, and silicon in an object that was born less than one billion years after the Big Bang. This distant quasar evidently contains the same quantities of these elements as much nearer quasars. That means that hydrogen and helium in the primordial gases had to be converted into heavier elements as early as one billion years after the Big Bang. And that raises the question of just how that could have happened. Was the first generation of stars already recycling elements back into the interstellar gas? Time will tell.

OPTIMIZING FOR COST

The Keck telescopes were bargains. Their innovative design cut the cost of construction by a factor of five, compared with a conventional telescope with the same area of glass. But even with the additional savings made by building two copies, they still cost over $70 million apiece. The Gemini telescopes cost even more.

Lawrence Ramsey, an astronomer at Pennsylvania State University, had a bright idea for a *really* cheap telescope. His specialty is spectroscopy, and he has invented several types of advanced fiberoptical spectrographs. Recently, he used one of these instruments in a search for global oscillations of stars, similar to those observed on the Sun. In the mid-1980s he began to think about a telescope devoted primarily to spectroscopy of faint objects. How could he design a telescope that smaller universities, like his own, could afford?

He recognized that the prime requirement for spectroscopy was plenty of light, which argued for a mirror at least 8 or 9 meters in diameter. He knew that the seeing at many locations was often no better than an arcsecond, which suggested that the mirror's angular resolution could be modest. And he had learned that most observations at mediocre sites were carried out within a relatively narrow part of the sky, near the zenith (the point overhead).

All these factors drew him toward the design of the Arecibo radio telescope. This 305-meter dish points straight up from a small valley in Puerto Rico. It is a transit telescope that tracks objects for a few hours as they pass from east to west. It can reach objects over a wide range of altitudes by moving the cluster of radio receivers around its prime focus. The dish is spherical in shape, not parabolic, a choice that reduced the cost of construction. Also, because the dish is fixed, no complicated mechanical mount is required.

Ramsey thought these principles could be applied to an optical telescope if one were willing to limit one's goals. He developed a design in collaboration with a number of colleagues and drummed up support for the project. In the early 1990s a consortium of universities was formed to build the Spectroscopic Survey Telescope. The group includes the universities of Göttingen, Munich, and Texas; Stanford University; and Pennsylvania State University. With adequate funds in hand, the consortium built the telescope at Fort Davis, Texas, the observing station of the University of Texas. It is one of the darkest sites in the continental United States and offers reasonably good seeing as well.

Figure 8.8 shows the final design of the telescope. By combining great light-gathering power with a multiobject spectrograph, this instrument is able to acquire the spectra of faint objects efficiently. The telescope is essentially a Gregorian, with a spherical primary mirror and a secondary that corrects for spherical aberration. The mirror is composed of ninety-one identical hexagonal glass segments, with an area equivalent to a 9.2-meter circular blank and

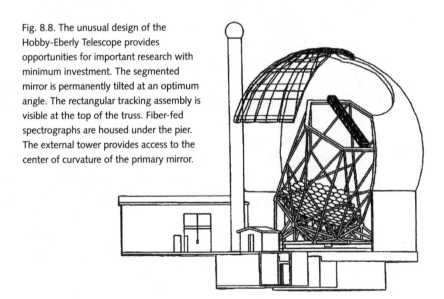

Fig. 8.8. The unusual design of the Hobby-Eberly Telescope provides opportunities for important research with minimum investment. The segmented mirror is permanently tilted at an optimum angle. The rectangular tracking assembly is visible at the top of the truss. Fiber-fed spectrographs are housed under the pier. The external tower provides access to the center of curvature of the primary mirror.

an effective focal ratio of f/1.8. The telescope points at a fixed altitude of 55 degrees above the horizon but rotates freely in azimuth about its vertical axis.

An array of actuators behind the mirror supports and aligns the segments. These actuators need to adjust the alignment of the segments only once during a night, because their orientation with respect to the force of gravity doesn't change during an observation. That circumstance greatly simplifies the design and cost of the actuators, in comparison with the Keck system, which has to adjust the alignment continuously.

This simplification comes at a price, however. To allow the telescope to track an object, the corrector optics and the spectrograph have to move in *six directions* on the spherical focal surface (north-south, east-west, in-out, and three angles of pitch, yaw, and roll). In effect the telescope has to track its target and maintain its focus simultaneously under computer control. With these motions, the telescope can reach over 70 percent of the sky visible from Fort Davis. Its field of view is relatively small, 4 arcminutes, but the optics are so fast that many such fields can be surveyed in one night.

The truss that holds the primary mirror was built in Germany to extraordinary specifications. It consists of an array of four hundred spherical nodes that connect the eighteen hundred spars of the structure. To make sure that the assembled truss would be rigid, each spar was machined to a precision of 1 micron.

The telescope was named the Hobby-Eberly Telescope (HET) to honor its principal sponsors, William P. Hobby (onetime governor of Texas) and Robert E. Eberly (a Pennsylvania philanthropist). It cost a mere $13.5 million, about 20 percent of other telescopes in the 8-meter class. It is now equipped with three spectrographs. A high- and a medium-resolution spectrograph are located in the basement and are fed with optical fibers. A low-resolution spectrograph rides on the corrector package.

The telescope was designed to survey large areas of the sky, rapidly and repeatedly. Some examples of its recent work include mapping the large-scale structure of the universe, at high redshifts; following up on transient phenomena such as supernovas and gamma-ray bursters; and determining the chemical composition of the Milky Way's halo. Several observing programs can be executed during a night by the resident astronomer, who has a queue of targets to work with.

The Hobby-Eberly Telescope began to observe in October 1999. Its immediate success alerted many astronomers at small institutions. If one could find a particular scientific niche and establish well-defined goals, one could build a large telescope at minimum cost. Another group of astronomers has done just that.

Eleven institutions in South Africa, Poland, Germany, the United States, New Zealand, and the United Kingdom are building a copy of the HET at Sutherland, South Africa. With its 11-meter aperture, the South African Large Telescope (SALT) will be the largest single telescope in the Southern Hemisphere. Like the HET, SALT will focus on spectroscopic surveys and monitoring transient events. SALT will cost its sponsors about $30 million and is scheduled for completion in December 2004. As of September 2003, the telescope mount and its enclosure were finished, and the critical tracking unit was being installed.

THE CANARIES SING LOUDER

As we've seen, the VLT, the Subaru, and the Gemini 8-meter telescopes all employ thin monolithic mirrors, with arrays of actuators to maintain their hyperbolic shapes. It begins to look as though 8 meters may be the practical limit of cast mirrors. Anyone with ambitions for more collecting area may have to follow Jerry Nelson's lead and build a segmented mirror or else assemble a cluster of 8-meter mirrors.

The Spanish Instituto de Astrofísica de Canarias has decided on the segmentation strategy. This major astronomical group is building a 10.4-meter copy of the Keck I telescope at the Roque de Los Muchachos, on La Palma. The Gran Telescopio Canarias (GTC), as it is called, is a classical Ritchey-Chrétien telescope. A third mirror directs light to any of three "folded Cassegrain" focal planes, as well as the prime, Cassegrain, and two Nasmyth foci. The primary mirror consists of thirty-six segments, 1 meter on a side, whose alignment, tip, and tilt are controlled continuously.

The scientific goals of the institute staff are appropriate for a telescope of this size. They include a search for extrasolar planets and brown dwarfs; star formation and proto-star structure; black hole studies; chemical abundances in distant galaxies; active and starburst galaxies; the early universe; and the properties of the intergalactic medium.

The Spanish government agreed to fund the full cost of the GTC if other

Fig. 8.9. Construction of the tube of the Gran Telescopio Canarias is continuing. This is a view inside the enclosure from summer 2004.

partners could not be found. So far the University of Florida and the Autonomous University of Mexico have each offered 5 percent of the construction and operating costs, in exchange for a corresponding share of the observing time.

Construction of the telescope is moving along on schedule. Schott Glaswerke delivered the Zerodur blanks for the thirty-six hexagonal mirror segments in December 2001. A French company, SAGEM, is stress-polishing the blanks. A Spanish engineering company finished building the telescope mount in March 2000. By October 2003 the dome was complete and installation of the telescope mount was in progress (fig. 8.9). In June 2004 the first segments of the mirror were being installed. At this rate, first light will probably occur in 2005. From then on, we can watch for some fabulous images to arrive from the Roque.

BEATING THE SEEING

WHY DO THE STARS TWINKLE? Most bright children ask that question sooner or later. Mom or Dad may reply that the air bends the starlight that reaches our eyes. Or they may ask the child to recall how distant cars on a hot black road seem to waver as we look at them.

These are pretty good answers. Experts know a good deal more about the way the atmosphere distorts our view of the universe, moment by moment. They talk about "atmospheric turbulence," but the basic idea is the same.

Astronomers have always had to live with this effect. They call it *bad seeing* because it ruins their observations. To avoid it, at least in part, they have located their telescopes at places where, for various reasons, the air is fairly stable and they can obtain sharp images. But even at good locations like La Silla or La Palma, the seeing can turn sour. A 4-meter telescope that could theoretically resolve details as small as one-twentieth of an arcsecond (its *diffraction limit*) will then be limited to perhaps 1 arcsecond (see note 5.2 on diffraction limit).

Astronomers have fumed about this situation for centuries and finally managed to do something about it. They have learned to build black boxes called *adaptive optics systems* that remove many of the disruptive effects of the atmosphere while they observe. Many telescopes around the world have been equipped with such hardware, at great expense and effort. These improvements have led to some interesting discoveries.

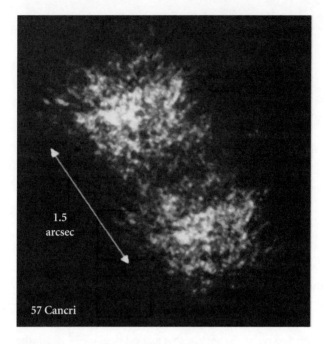

1.5
arcsec

57 Cancri

Fig. 9.1. Speckle images of two stars in a "wide" binary system. Notice that the patterns are similar. Each speckle is a diffraction-limited image.

SPECKLES

To learn how to build adaptive optics, we had first to understand exactly how a telescope forms an image through a messy atmosphere. Antoine Labeyrie was one of the first to see into the problem. He is a French astrophysicist, born in 1943, who spent many years at the Nice Observatory and is now a full professor at the Collège de France in Paris.

In 1970 he pointed out a curious phenomenon. If you look at a star through a telescope with an ordinary eyepiece, the image appears as a shimmering blob of light a few arcseconds in width. With sufficiently high magnification, however, one can see that the blob consists of a cloud of tiny "speckles" of light that dance around furiously (fig. 9.1). Labeyrie realized that each speckle is a faint, diffraction-limited image of the star (i.e., an image with the resolution the telescope would have if it were located above the atmosphere).

How are the speckles formed? Figure 9.2 may help us to understand. Light from a star moves out in a parade of spherical wave fronts, on which the phase of the light's vibration is a constant. A wave front passes through cells of warmer and cooler air, a few centimeters in size, on its way to the telescope. The cells behave like weak lenses and tilt parts of the wave front in different directions. Or if you prefer to think of it differently, the cells retard some parts of the wave front relative to others. When the wave front arrives at the telescope, it is no longer spherical but wrinkled. The air bends infrared light less than visible light, so the wrinkles are longer and flatter for infrared light.

All the parts of a distorted wave front with the same slope are combined by the telescope mirror into a single faint speckle not much bigger than the diffraction limit of the telescope (note 9.1). The number of speckles is roughly equal to the number of air cells that could fit into the area of the mirror. So if a typical cell size were 10 cm, a stellar image could have as many as twelve hundred speckles at any instant at the focus of a 4-meter telescope. And because the air cells are swept along in a high-altitude wind, the pattern of speckles changes markedly within a few milliseconds.

In a long photographic exposure, all the speckles blur together into a blob a few arcseconds in size, much larger than the telescope's diffraction limit. In 1970 Labeyrie recommended taking hundreds of very short exposures that freeze the motion of the speckles, to get a good sample of the atmosphere's behavior. Then he showed how to use such exposures to find the separation of the stars in close binary systems.

Suppose that the angular separation of the stars, as seen from the telescope, is smaller than the angular width of a typical cell, or *isoplanatic angle*. Then the cell will bend the wave front of each star by the same amount. In fact, *every* cell within the telescope's field of view will bend the two wave fronts equally,

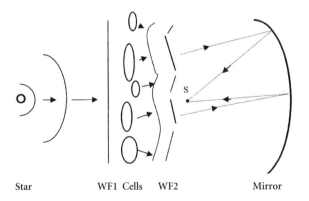

Star   WF1 Cells   WF2   Mirror

Fig. 9.2. Speckles are formed as light from a source passes through a layer of cool and warm air cells. Each cell tilts a part of the incoming wave front. The telescope collects all parts of the wave front with the same slope and forms a speckle at its focal plane.

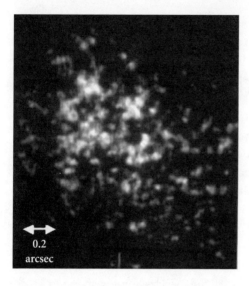

0.2
arcsec

Fig. 9.3. In a short exposure, the speckle patterns of two stars in a close binary overlap but still look similar.

but different cells will bend them by different amounts. That means that the two speckle patterns will be identical in a short exposure. But because the stars' separation is less than a cell's width, the speckle patterns overlap (fig. 9.3).

At first glance the problem of finding the separation might seem hopeless. But Labeyrie showed how to disentangle the two speckle patterns by *autocorrelating* them. Basically, that means sliding one pattern over a copy of the other, until the two patterns match. The distance you need to slide the two patterns is then the separation of the stars. When Labeyrie was finished, he had obtained not only the separation but also the size of a typical speckle. That is the image of a star the telescope would obtain outside the atmosphere and is broadened only by the aberrations of the telescope mirror.

Binary stars are the most important sources of stellar masses, which are essential for an understanding of star formation. Harold McAllister and his group at Georgia State University have been using Labeyrie's technique for twenty years to obtain the orbits, masses, and luminosities of thousands of binary stars.

THE NEXT STEP

Speckle interferometry, as it is called, works best for simple objects like multiple star systems. And for it to work at all, the light from the source should be limited to a narrow band of wavelengths so that it will interfere with itself. That means discarding precious photons, a practice that astronomers frown on. So this technique has its problems.

But Labeyrie's insight led to further developments. One can think of an extended object, like a galaxy or a planet, as consisting of many point sources, each of which is affected by the atmosphere in the same way as a star. That leads us to the possibility of sharpening the image of a broad object by using the speckle pattern of a nearby star as a guide.

Betelgeuse is such a broad object. It is a huge red star in the constellation Orion. In 1922 Albert Michelson and Francis Pease measured its angular diameter with a stellar interferometer (note 9.2) on the Mount Wilson 100-inch (2.5-meter) telescope. Its linear diameter turns out to be about the size of the Earth's orbit around the Sun, making it a suitable target for Labeyrie's method. In 1976 Roger Lynds, Peter Worden, and John Harvey used speckle interferometry at the Kitt Peak National Observatory 4-meter telescope to reconstruct a crude image of Betelgeuse's disk. This was the first time the surface of a star other than the Sun had been seen.

To pull off this trick, they first measured the speckles of Gamma Orionis, a pointlike star, to determine the blurring effect of the atmosphere. With that information they were able to process the speckles of Betelgeuse. They were lucky, because Gamma Orionis is several degrees distant from Betelgeuse. As a general rule, one needs a pointlike star within a few tens of arcseconds of one's actual target to ensure that the seeing is the same for both objects.

For some extended objects, like the Sun or a faint galaxy, no suitable reference star is available. Is it possible to restore a blurred image of such a target? In a landmark paper of 1974, K. T. Knox and B. J. Thompson (University of Rochester) proposed a mathematical procedure to accomplish that task. Labeyrie's method recovers only the *brightness* of the target but loses all information on its detailed structure. For a full reconstruction, one needs not only the brightness but the *phase* of the wave front that each point in the object emits. Knox and Thompson demonstrated their method with a one-dimensional computer simulation, but their method is quite general. It soon became popular with astronomers everywhere.

In 1983, for example, Robert Stachnik, Peter Niesenson, and Robert Noyes (Smithsonian Astrophysical Observatory) applied the Knox-Thompson procedure to sharpen images of the Sun. They first obtained a long series of exposures in a narrow wavelength band at the 1.5-meter solar telescope at Kitt

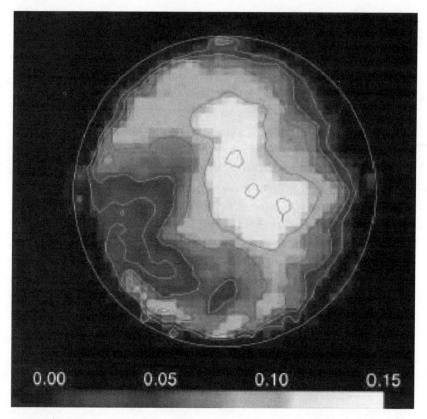

Fig. 9.4. This 2-micron image of Titan, Saturn's largest satellite, was obtained with speckle interferometry at the Keck II telescope in 1999. Titan has an angular diameter of 0.7 arcsecond and a linear diameter of 5,150 km. The Cassini probe passed close to Titan on July 2, 2004, and obtained higher-resolution images of the cloudy surface.

Peak. Then they processed them with the Knox-Thompson recipe. The final images of sunspots revealed filamentary details as small as 0.11 arcsecond, ten times smaller than the seeing would allow otherwise.

In 1999, astronomers at the Keck Observatory obtained speckle images of Saturn's largest satellite, Titan, whose true diameter is only 0.7 arcsecond. They processed eight hundred images to obtain one beauty, with a resolution of 0.04 arcsecond, the diffraction limit at a wavelength of 2.2 microns. For the first time they could see surface details (fig. 9.4). They suggested that the dark regions might be seas of liquid hydrocarbons and bright regions could be highlands of ice and rock.

ADAPTIVE OPTICS

Speckle interferometry is a powerful tool, but it requires a lot of processing on a lot of data after a session at the telescope is over. In 1953 Horace Babcock, a solar astronomer at the Mount Wilson Observatory, got an idea for a gadget that might remove the effects of the atmosphere on an image of the Sun *during the observation*. He proposed to use an electron beam to vary the thickness of a liquid film that was spread over a fixed glass plate. Thicker parts of the film would slow down those parts of a wave front that were ahead of the others. That would smooth the wrinkles in the wave front. It was a good idea but too advanced for the contemporary technology and was not followed up.

Robert Leighton, Babcock's colleague at Mount Wilson, had a more practical idea. He used a small mirror that could tilt (left and right) or tip (forth and back) very rapidly in response to a signal from a simple detector at the edge of an image of a planet. The mirror removed the slow drifting of the image in the focal plane, which is one cause of poor image quality, but did little to remove the blur.

Adaptive optics (AO) as we know it today really got started in the 1970s, when the U.S. Defense Department faced a new challenge. The Soviets were launching many military satellites, and the department wanted to be able to take a close look at them in sunlight. But the turbulent atmosphere frustrated their attempts to get clear images. A research program was hastily organized to develop tools to obtain clear images continuously during an observation.

John W. Hardy was a member of a team of physicists at Litton Itek Optical Systems that was hired to solve the problem. The work was classified, of course, and was released to the public only after the Berlin Wall came down and the Soviet Empire collapsed. In the June 1994 issue of *Scientific American* Hardy relates how the team struggled to build a practical system.

The task was horrendous. The targets were not simple pointlike stars but extended objects. The distorting effects of hundreds of air cells in a telescope's field of view had to be removed to get a sharp image. The job had to be accomplished within a few milliseconds, before the cells changed, and then repeated, over and over. What's more, there was relatively little light to work with, and yet the final wave fronts had to be spherical to within one-tenth of the wavelength of light.

Hardy's team broke the problem down into three parts: *measuring* the

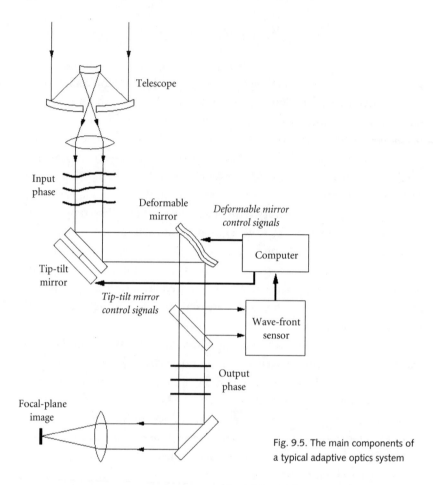

Fig. 9.5. The main components of a typical adaptive optics system

wave-front distortion; *calculating* a correction; and *applying* the correction to the wave front. Each task would be attacked with a specific piece of hardware (fig. 9.5). This basic architecture has been adopted by astronomers and physicists ever since.

If the team could see a pointlike star within a few arcseconds of a satellite, they could use its blurred image as a reference with which to correct the image of the satellite. Hardy and his team were not so lucky, however. They couldn't be sure to have a star in view, so they had to use the image of a satellite as its own reference.

The solution they found was the *shearing interferometer*. With this device, well known to optical scientists, two copies of the same monochromatic image are superposed with a known separation. The images interfere to produce a pattern of light and dark "fringes." The intensities of the fringes are measures of the *slopes* of the wrinkles on a wave front. This only works in monochromatic light, however. Hardy's team had to utilize *all* the light from a faint satellite, so they invented a shearing interferometer that works in white light. That was no mean feat.

These measurements of the slopes of a wrinkled wave front were sent to a *reconstructor,* a black box that calculated the phase corrections that had to be applied for each air cell in the field of view. Finally, these phase differences, or, equivalently, path differences, were sent to a small deformable mirror. The mirror adjusted the path lengths of all parts of the wave front so that they would arrive at the telescope mirror in phase. In short, it ironed out the wrinkles in the wave front.

Hardy's deformable mirror was designed to correct the phase of only twenty-one segments of a wave front. It consisted of a thin glass plate mounted on a block of piezoelectric material. An electrode was provided for each of the twenty-one segments. When the reconstructor applied a voltage to an electrode, a piezoelectric actuator expanded by the distance needed to equalize the path length to the telescope mirror. The mirror was placed in the optical path of the telescope, ahead of the final focal plane, so that a photograph or digital record could be made of a sharpened image.

A prototype of this Real-Time Atmospheric Compensation system was completed in 1973. Despite the small number of wave-front segments the system could handle, it demonstrated that the method worked (note 9.3). The device could operate reliably thousands of times per second to keep up with the changes in the seeing. Hardy's group went on to build a system with 168 electrodes for a 1.6-meter telescope, operated by the U.S. Air Force.

ANOTHER PIONEER

The military was far ahead of the astronomers in building practical systems to eliminate bad seeing. Indeed, hardly anyone else could afford to build such a system in the 1970s, even if they had the ideas. But physicists and astronomers were active in developing a theory of atmospheric turbulence that would prove to be essential to further development. As early as 1966 David Fried showed how to estimate the average size of the wrinkles on a wave front using basic parameters of the atmosphere. More precisely, he determined the angle within which the phase of the wave front was nearly uniform. This angle he called the *isoplanatic angle.*

Fried learned that the isoplanatic angle depends on the wavelength in which you look. He assumed he could replace the effects of a deep atmosphere by a thin layer, at a typical altitude of about 20 km. Then at visible wavelengths, the linear width of a wrinkle (r_0) is about 10 cm, while in the infrared (say, at 2 microns), r_0 increases to 50 cm. That means that an infrared wave front has fewer wrinkles across the diameter of a telescope's primary mirror. An AO system that corrects infrared images would therefore be simpler and cheaper to build.

Several clever physicists were also proposing hardware to obtain sharp images, well before the military systems were declassified. For example, a team at the University of California at Berkeley, led by Andrew Buffington, experimented in 1976 with six movable mirrors to correct an image. But Robert Smithson, a student of Robert Leighton at Caltech, was one of the first to cobble together a complete system, with support from the Lockheed Research Laboratories. His system was designed to correct only nineteen wrinkles on a wave front, but this was quite sufficient for a proof of the concept.

Smithson's deformable mirror contained nineteen hexagonal segments arranged in a circular array 15 cm in diameter. Each segment could be tilted and tipped by three piezoelectric pistons or actuators. His wave-front sensor was adapted from a device invented by Jacob Hartmann in 1901 to measure the aberrations of optics. This *Shack-Hartmann array* consisted of nineteen little lenses, one for each segment of his mirror (fig. 9.6). Each lens formed an image of the target on a charge-coupled device. If the seeing was perfect, each image would lie on the axis of the lens. If the seeing was bad, however, the images would be displaced laterally. Each displacement is a measure of the slope of a segment of the wave front.

The Shack-Hartmann sensor has two big advantages. It is simple to build and maintain, and, more important, it works with white light. Unlike the standard shearing interferometer, it utilizes all the available light from a target. That ability is crucial for astronomers.

In 1983 Smithson tested his system on images of the Sun on the 0.76-meter solar telescope (the Vacuum Tower Telescope, later rededicated as the Dunn

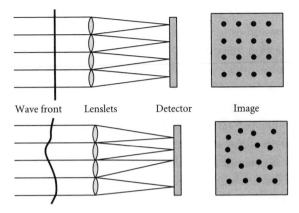

Wave front   Lenslets   Detector   Image

Fig. 9.6. The Shack-Hartmann sensor is an array of little lenses that form images of a distant source on a photon detector. Without an atmosphere, all the images would lie in a perfect square array, but with bad seeing, the images are displaced. Each displacement measures the slope of a part of the wave front.

Solar Telescope) at the Sacramento Peak Observatory in Sunspot, New Mexico. Obviously he couldn't use a pointlike star to measure the distortions of the atmosphere, so he used the sharp borders of dark sunspots instead. Sunspots are embedded in the *granulation,* a pattern of bright convection cells that are separated by darker lanes. The overall effect is that of a network or mesh. Smithson's system produced sharper images of the granulation even in moderately poor seeing. That success demonstrated the power of adaptive optics, at least on a bright object like the Sun. (Note 9.4 describes the latest in solar AO.)

THE FIRST STELLAR ADAPTIVE OPTICS SYSTEM

As we learned earlier, the members of the European Southern Observatory decided in 1987 to build the Very Large Telescope at Cerro Paranal. From the first they planned to equip their four 8-meter telescopes with an adaptive optics system. In the late 1980s enough information was available in the open literature about the components of such systems to allow them to proceed rapidly.

A team of French astronomers and engineers, including Gerard Rousset and Fritz Merkle, collaborated with a French firm, Laserdot, and put together a working system in just a couple of years. They called this prototype Come-On (*Astronomy and Astrophysics* 230, L29, 1991). As in Labeyrie's speckle-imaging technique, this system required a pointlike star close to the actual target in order to correct the wave fronts.

Come-On had the usual three components: a Shack-Hartmann wave-front sensor, a deformable mirror, and a reconstructor. In addition, a separate tip-tilt mirror was provided to remove the slow motions of the whole image. The Shack-Hartmann sensor measured the wave-front distortions in visible light. It contained twenty-five little lenses in a square array, one lens for each segment of a wave front. The deformable mirror was a special design. Unlike Smithson's mirror, it didn't have discrete segments. Instead, it consisted of a continuous reflecting layer, bonded to an array of nineteen piezoelectric actuators. When a predetermined voltage was applied to an actuator, it expanded by a precise amount and changed the local curvature of the mirror.

Come-On used two coupled computers: one to synchronize the components of the system and the other to calculate a set of wave-front corrections every ten milliseconds. (See note 9.5 for a description of how a generic reconstructor calculates these adjustments.)

The final image was presented to a digital camera. Only the infrared light of the image was recorded, however. In that way the system was taking advantage of Fried's discovery that infrared wave fronts are less affected by the atmosphere.

In 1989 the whole rig was tested at the 1.5-meter telescope at the Observatory of Haute-Provence, France. It produced the first diffraction-limited images of a faint astronomical object ever made with adaptive optics, a major achievement. The members of the binary system Gamma2 Andromedae are only 0.5 arcsecond apart and were resolved at a wavelength of 2.3 microns in seeing that varied between one and two arcseconds. A pointlike star, Gamma1 Andromedae, at a distance of 9.6 arcseconds, was used to sharpen the image of the binary.

A sharper image is also a brighter one. In the tests at Haute-Provence, for example, the AO system increased the central brightness of the stars' images by a factor of twenty. This is another big advantage of adaptive optics.

A second-generation system called the Adaptive Optics Near-Infrared System, or ADONIS, was built with a forty-nine-element sensor and shipped to ESO's La Silla Observatory in Chile. There it was installed for general use at the 3.6-meter. ESO was well under way toward its goal of an advanced system for the VLT.

CURVATURE SENSING: AN ELEGANT SOLUTION
TO A DIFFICULT PROBLEM

François Roddier is one of the most inventive players in the game of optical interferometry. He graduated with a doctorate in physics from the University of Paris in 1964 and settled at the University of Nice. There he helped to develop a novel device (an atomic beam spectrometer) for solar velocity measurements. He and his wife, Claude, became interested in the physics of seeing in the mid-1970s, and that led them into the problems of sharpening astronomical images. Since the 1980s the Roddiers have worked on essentially every aspect of astronomical imaging and interferometry and have made many fundamental contributions.

In 1985 the Roddiers moved to Hawaii and joined the Institute for Astronomy. Three years later François published what may be his most influential idea, curvature wave-front sensing. Figure 9.7 demonstrates the principle. Here we see an incoming wave front near the focal plane of a telescope. As it passes through, its brightness distribution is measured on two planes, one in front and the other behind the exact focal plane of the telescope. If the wave front has a bump, the front plane will be fainter and the back plane will be brighter at the location of the bump. The difference of the light intensities, divided by their sum, yields the *curvature* of the bump (i.e., the *slope of the slope* of the wave front, in the vicinity of the bump). This quantity contains more information on the shape of the wave front than a simple measurement of the height of the bump.

In addition, the overall tilt and tip of the wave front can be determined from measurements at the edges of the field of view. So the brightness distribution over these two planes yields all the information needed to reconstruct the wave front. One nice feature is that this sensor works well in white light and so is able to utilize all the light from a faint source. It does require a pointlike star close to the actual astronomical target, however.

The sensor was recognized as a clever innovation. But its true power and elegance emerged only after Roddier discovered a way to reconstruct the wave front without the use of a reconstructor, the black box that calculates the corrections. He mated his sensor with a special type of deformable mirror, a *bimorph*. Bimorphs are used as the active element in an audio tweeter and in other industrial applications. A bimorph consists of two thin piezoelectric

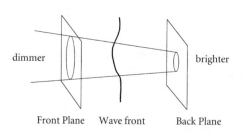

dimmer   brighter

Front Plane   Wave front   Back Plane

Fig. 9.7. In a curvature sensor the brightness of a wave front is measured on two planes, one in front and one behind the exact focal plane of the telescope. The difference of brightness at corresponding points on the planes is a measure of the gradient of the slopes of the wave front.

sheets that are bonded together. When voltages are applied to the bimorph's electrodes, one sheet expands and the other contracts, causing the pair to bend into a precisely determined curve.

In the late 1980s Roddier and his team at the University of Hawaii built a thirteen-electrode prototype of a complete curvature system (note 9.6). In August 1995 the system was used at the Canada-France-Hawaii Telescope on Mauna Kea to observe Saturn at the moment that its rings appeared edge-on. The system achieved a resolution of 0.15 arcsecond in the near infrared, which allowed Roddier and friends to detect several faint satellites previously seen only from the Hubble Space Telescope. Clouds on Neptune were also resolved for the first time.

The thirteen-electrode prototype was so successful that a team at the CFHT began to build a slightly larger copy. This nineteen-element system was called Pueo, which means "little owl" in the Hawaiian language. Pueo was commissioned on the CFHT 3.6-meter in 1996 and later installed on the 3.6-meter United Kingdom Infra Red Telescope. Pueo has since been used to monitor the variations in cloud structure of Neptune with a resolution of 0.12 arcsecond in the infrared.

In 1997 Pueo's big brother, Hokupa'a (immovable star), was completed. It has thirty-six segments in its sensor and bimorph, and is therefore better suited to larger telescopes. It was first tested on the CFHT and in 1999 was mounted at the 8-meter Gemini North telescope. The Gemini consortium was so pleased with its performance that it decided to build a 185-segment version called ALTAIR. During its first tests in June 2003, ALTAIR achieved a resolution of 0.06 arcsecond at a wavelength of 1.6 microns. That's the width of a housefly at a distance of 20 km.

ARTIFICIAL STARS

To do its job, an AO system needs a pointlike star that lies close enough to the actual target to undergo the same amount of blurring (i.e., the star must lie within the same isoplanatic angle as the target). Finding such a *guide star* is often a problem, for two reasons. First, as every sky watcher knows, one sees fewer stars as one looks farther from the plane of the Milky Way. Second, a guide star must be bright enough for an AO system to capture sufficient light in only a few milliseconds.

François Roddier has estimated the probability of finding a suitable guide star at a moderate distance (30 degrees) from the Milky Way's plane. If one observes in visible light (say, a wavelength of 0.63 micron) to find the star, the probability is less than 1 percent, because the isoplanatic angle is only 13 arcseconds and because the star must be brighter than magnitude 13 (note 9.7).

If that were the final answer, adaptive optics would be useful for only a small fraction of desirable objects. However, everything improves in the infrared, as David Fried pointed out long ago. The chance of finding a guide star using a wavelength of 2.2 microns rises to 90 percent, because the isoplanatic angle is now 58 arcseconds and the guide star can be as faint as magnitude 17. But even at 2.2 microns, the chance of finding a suitable guide star at the galactic pole is only 50 percent.

What can one do? Antoine Labeyrie had already suggested a solution to this problem. In 1985, just as astronomical adaptive systems were beginning to be planned, Labeyrie and his colleague Roger Foy realized that one should be able to manufacture artificial stars with a laser. They proposed two different schemes. If one sent pulses of blue laser light toward the target, the atmosphere would scatter light from the beam back to the telescope. (This *Rayleigh scattering* is responsible for the blue color of the sky.) A fast shutter could be timed to open only to the light returning from a chosen altitude. The laser spot at that altitude would look like a star to an adaptive optics system.

An even better scheme, Labeyrie and Foy said, was to tune a laser to a powerful spectral line (589 nm) of sodium. (This is the yellow light emitted by sodium vapor lamps on highways.) A French group, using lasers to probe the atmosphere, had detected a layer of meteoritic sodium atoms at an altitude of about 95 km. Labeyrie and Foy calculated the size and brightness of the spot a sodium laser could produce in this layer. Their results were encouraging. The scheme seemed quite feasible.

Laird Thomson, an astronomer at the University of Hawaii's Institute for Astronomy, pounced on the idea. He contacted Chester Gardner, at the University of Illinois, who was studying the upper atmosphere with lasers. This type of work was called LIDAR, an acronym for Light Detection and Ranging, the optical analog to RADAR. In January 1987 Thomson, Gardner, and two graduate students installed a laser at the 0.6-meter telescope of the University of Illinois, just 50 meters from Hawaii's 2.3-meter telescope on Mauna Kea. The laser shot a yellow beam skyward, and the 2.3-meter telescope collected the light scattered from the spot in the sodium layer. A bright yellow blob appeared, as expected. Astronomers would have to learn to focus the light of lasers before the method would be useful, but the principle was established.

Scientists at the Starfire Optical Range, an Air Force facility near Albuquerque, were surprised to hear about this experiment. As part of the classified Star Wars project to shoot down enemy missiles, this group had already built adaptive optics systems, complete with laser guide stars. They may have been aware of Labeyrie's ideas or conceived them independently. In any case, they had demonstrated that both Rayleigh scattering stars and sodium layer stars worked effectively.

After Thompson and Gardner published their work on sodium stars, there was less reason to keep the Air Force work classified. Around 1990, after the Star Wars project was terminated, much of the secret work on adaptive optics and laser stars was released to the public. Astronomers were then able to advance rapidly in building their AO systems.

LICK OBSERVATORY SCORES A FIRST

Lick Observatory was the first to operate an AO system with a laser guide star. The project proceeded in steps over several years. First, Lick astronomers collaborated with physicists at the Lawrence Livermore National Laboratory in building a prototype AO system. It employed a continuous-faceplate deformable mirror with sixty-nine actuators and a Shack-Hartmann sensor. The first trials of this early system were made in 1993 on Lick's 1-meter Nickel Telescope, using natural guide stars. Then in 1995 a second system with 127 actuators was tested on the 3-meter Shane Telescope. It worked beautifully, shrinking the raw image of a star by a factor of six.

Next, an auxiliary sodium laser facility was set up outside the dome of the Shane Telescope. The pulsed laser produced a spot about 1 meter in diameter in the sodium layer at an altitude of 90 km. These tests, conducted in 1995, went so well that a permanent facility was attached to the 3-meter telescope. The lasers are located in the dome's basement, and their light is piped up to the telescope with fiberoptics. Thus the 3-meter acts as both the transmitter of the laser pulse and the receiver of the laser light.

In October 1996 the new system was tested on a pointlike star. The peak intensity of the corrected image was increased by a factor of 2.4, and the width of the star's image was reduced from 0.75 to 0.31 arcsecond. These first results have since been significantly improved.

The Lick system has been used routinely since 1996. Moreover, it has served as a test bed for an AO system, with laser guide stars, for the Keck II 10-meter telescope on Mauna Kea.

The laser guide star story has an ironic twist. An AO system needs a *natural* guide star, in addition to the laser star, to function properly. One reason is that the laser beam crosses the atmosphere twice (once going up and once coming down), while the light from the astronomical target crosses it only once. Therefore, the tilt of the laser star's wave front cancels to zero, but the tilt of the target's does not. A natural star must supply the proper tip-tilt correction. So although laser guide stars were invented to avoid using natural stars, they are handicapped without them.

Moreover, sodium laser guide stars are not totally reliable. The sodium layer in the atmosphere is maintained by in-falling meteors, so the density of sodium atoms depends on the frequency of meteor showers, and also on seasonal factors. The layer can vary appreciably within a few hours, during a long astronomical exposure. Nevertheless, laser guide systems are enormously helpful, and many observatories have installed them.

EVERYBODY'S GOT THEM

In the 1990s, adaptive optics took off. François Rigaut, an astronomer at the CFHT, listed twenty observatories around the world that had either a working system in 1996 or serious plans for one. They were installed on telescopes with mirrors as small as 1.5 meters and as large as the Palomar 200-inch (5-meter). Many of them were given imaginative names and pronounceable acronyms. All of them were designed to produce diffraction-limited images at near-infrared wavelengths between 1 and 5 microns. Here are a few examples.

At Mount Wilson, the 100-inch (2.5-meter) Hooker Telescope was equipped in 1995 with ADOPT, an 8 × 8 element Shack-Hartmann sensor. The 3.9-meter Anglo-Australian Telescope at Siding Springs was fitted with a curvature AO system in 1996, and the Palomar 200-inch (5-meter) got a Shack-Hartmann system called PALAO in 1999.

At La Palma, the William Herschel 4.2-meter was introduced to NAOMI in 1998, and Durham University's telescope was blessed with ELECTRA. The 3.5-meter at Calar Alto, in Spain, received a system that was christened ALFA.

CHAOS is the name of the AO system for the 3.5-meter telescope at Apache Point, New Mexico. It is far from chaotic, in fact, and works nicely, especially with its new laser guide star system. It incorporates a 16 × 16 Shack-Hartmann sensor and a home-built continuous-faceplate mirror with 201 actuators.

By 2000, the 8- and 10-meter telescopes on Mauna Kea were being equipped with large, second-generation systems, all with plans for laser guide stars. These gadgets are not cheap. The AO system for the Keck II telescope, for example, was funded with $6.3 million from the Keck Foundation and $1.2 million from Caltech. It consists of a 64 × 64 element Shack-Hartmann sensor and a 349-actuator continuous-faceplate mirror built by the Xinetics company.

The Japanese chose to build a curvature AO system for Subaru, their 8.3-meter national telescope on Mauna Kea. It has a thirty-six-actuator bimorph mirror and a separate tip-tilt mirror. ALTAIR, at the Gemini North, contains a Shack-Hartmann sensor with 12 × 12 elements and a 177-actuator deformable mirror.

The European Southern Observatory has an especially large agenda for the four units of the Very Large Telescope. ESO has built NAOS, a Shack-Hartmann plus 185-actuator deformable mirror combination for a Nasmyth port on Yepun, the Unit 4 telescope. SINFONI is a sixty-element curvature system for Yepun's Cassegrain focus. Both of these AO systems are intended for use with infrared spectrographs.

MACAO, an acronym for Multi-Application Curvature Adaptive Optics, is another sixty-element system. Each of the four units of the VLT will be equipped with a copy of MACAO when the VLT operates as an optical inter-ferometer. We'll have more to say about that in the next chapter.

ASTRONOMY IN DAYLIGHT

Solar astronomers have a special problem. They don't usually want for light, but resolving the menagerie of tiny magnetic and dynamic structures in the Sun's atmosphere has always been a struggle. As an example, consider figure 9.8, which shows the fine details in the border of a sunspot. These filamentary

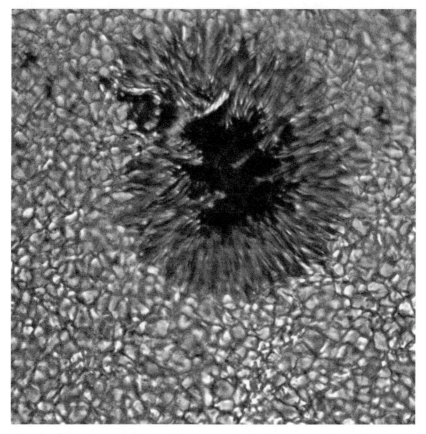

Fig. 9.8. Fibrils in the penumbra of a sunspot are shaped by the magnetic field lines that spread from the center. The sunspot lies in a field of solar granules, each about the size of Alaska. Their sub-arcsecond dark borders can serve in place of a point source for an adaptive optics system.

structures are no wider than a few tenths of an arcsecond. They are shaped by the magnetic field lines that spread out from the sunspot. To understand how sunspots form and decay, solar physicists want to study such features for hours. The daytime seeing at most sites rarely is good enough or lasts long enough, however.

Although most of the turbulence that causes bad daytime seeing lies above the telescope, a fair amount resides within the telescope. To produce a conveniently large solar image, say a meter across, a solar telescope must have a long focal length. The McMath-Pierce Solar Telescope on Kitt Peak, for example (see fig. 4.1), has an air path 83 meters long. Unless the air in the telescope is perfectly stable, however, its turbulence will blur the solar image. One solution, pioneered by Richard Dunn at the Sacramento Peak Observatory, was to evacuate the light path to the main mirror. His 0.76-meter Vacuum Tower Telescope achieves a resolution of a few tenths of an arcsecond, but generally for only an hour or so in the morning. Solar astronomers have used the VTT for three decades to make some fundamental discoveries. To make further progress in understanding how the Sun works, however, they need even more resolution.

The answer is, of course, adaptive optics. But an AO system needs a point-like reference target, like a natural star or laser guide star. Neither of these is available in broad daylight. The *solar granulation,* a pattern of convection cells near the solar surface, offers the best substitute (see fig. 9.8). The dark lanes between the cells, a few tenths of an arcsecond in width, could act as a resolution test for an AO system.

Developing an AO system for a solar telescope has been a long, hard struggle. Scientists at the Lockheed Palo Alto Laboratories collaborated with Richard Dunn and his colleague Thomas Rimmele to build a series of prototypes. Astronomers at the Kiepenheuer Institute for Solar Physics (Freiburg, Germany) developed their own design. The latest version at the VTT is a Shack-Hartmann system with seventy-six apertures. It routinely produces quarter-arcsecond resolution.

This system is seen as the test bed for a system that would work on the 4-meter Advanced Technology Solar Telescope, which the U.S. National Solar Observatory is designing. The goal is to achieve a resolution of 0.03 arcsecond in visible light, or a factor of ten better than the VTT obtains.

CENTER, ANYONE?

As you can see, many observatories have established a home team of experts to design and build adaptive optics systems for their telescopes. The teams sometimes collaborate with another observatory or an industrial firm. In addition , several specialized centers have been established to invent new systems and to adapt older designs to the monster telescopes that are being planned. (We'll get to those further on.)

For example, Jerry Nelson now directs the Center for Adaptive Optics (CFAO) at the University of California at Santa Cruz. This group is designing the AO hardware for a 30-meter telescope, the California Extremely Large Telescope (CELT).

Similarly, Roger Angel heads the Center for Astronomical Adaptive Optics at the University of Arizona in Tucson. His group is working on a black box for the Large Binocular Telescope and for the 6.5-meter conversion of the Multi-Mirror Telescope, among other projects.

Adaptive optics has some surprising applications outside of the field of astronomy. At the Center for Visual Science at the University of Rochester and the College of Optometry at the University of Houston, scientists are using AO to correct optical imperfections of the human eye. They can also study the organization and function of the eye's retina, for clinical applications. And the University of Moscow has demonstrated the use of AO to improve high-power laser welding and cutting of metals. Who knows what else these amazing gadgets will do?

SCIENCE HIGHLIGHTS

Solar system objects were some of the first targets astronomers reexamined with AO. The clouds on Neptune (fig. 9.9), the rings around Uranus (fig. 9.10), the minor satellites of Saturn, and the asteroid Kleopatra were all popular.

Io, one of the four Galilean satellites of Jupiter, is also a favorite (fig. 9.11). Jupiter's tidal forces on Io generate internal heat, which occasionally emerges in violent volcanic eruptions. Several telescopes with AO systems, including ADONIS on the 3.6-meter at La Silla, have been monitoring Io's activity for a number of years. In 2002 a large team at the Keck II witnessed two eruptions

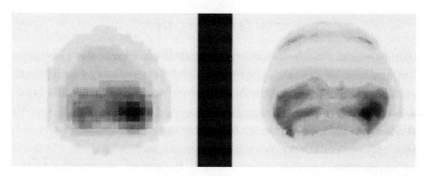

Fig. 9.9. Images of Neptune, obtained at the Keck II telescope, with *(right)* and without adaptive optics operating

on Io with their new AO system. They were able to determine the thermal output of one of these, which turned out to be the largest ever seen.

The central regions of quasars and other active galaxies are being studied intensively with adaptive optics to learn more about the mechanisms that produce their extraordinary emissions. In 2003 Gabriela Canalizo and her colleagues used the Keck II AO system to study the central core of the powerful radio galaxy Cygnus A. They found that they could not resolve its tiny nucleus, even with a resolution of 0.05 arcsecond. But within 0.4 arcsecond of this nucleus they discovered a point source, most likely the dense core of a satellite galaxy that is being devoured by the giant elliptical galaxy. They suggested that the merger of these two objects is powering part of the radio emission.

Brown dwarfs are stars with very low masses and luminosities. Less is known about them than other classes of stars because they are so faint. Nevertheless, it has been suggested that they contain at least some of the missing mass, or dark matter, in the universe (see note 4.3). Several observatories are searching the heavens to determine how numerous they are. Other astronomers are interested in their intrinsic properties and are using AO in conjunction with spectroscopy to study them.

A rare binary star system, consisting of two brown dwarfs only 0.13 arcsecond apart, was discovered with the Gemini AO system in 2001 by a team from the University of Hawaii. A year later a team at the Subaru telescope examined the pair spectroscopically with their natural guide star AO system and an infrared camera. The spectrum of one star contains multiple spectral lines of water vapor, which pegs the temperature of the star's surface at less than 1,500 kelvin. (For comparison, the Sun's surface temperature is about 5,700 kelvin.)

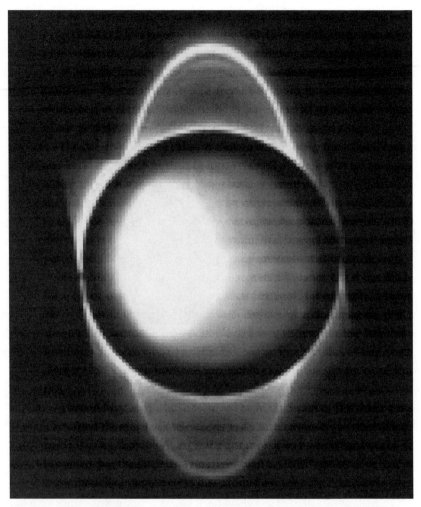

Fig. 9.10. This diffraction-limited infrared image of Uranus and its rings was obtained by the Keck II AO system. Uranus has an angular diameter of about 3.6 arcseconds, so the width of the bright ring is about 0.1 arcsecond, or 1,500 km. As it flew by Uranus, the Voyager satellite discovered ten rings, varying in width from 100 km to just a few kilometers.

Fig. 9.11. A 2.2-micron image of Io, one of the four Galilean moons of Jupiter. It was obtained with the adaptive optics system NAOS-CONICA on the Very Large Telescope in December 2001. Io has a diameter of only 1.2 arcseconds, but the system resolves features 0.063 arcsecond in size.

Such observations help to reveal the physical properties of the atmospheres of such faint stars.

The search for life elsewhere in the universe has become a major objective for astronomers. Mars is a prime target, and in 2004 both the British and the Americans sent robot rovers there to look for water and, possibly, life. (Both American rovers did find signs of ancient lakes.) At the same time, a vigorous search for inhabitable planets around other stars is under way. As of 2003, over a hundred Jupiter-sized planets have been found, orbiting stars other than our Sun. Many of these were discovered by the California and Carnegie Planet Search team, which includes R. Paul Butler, Geoffrey Marcy, and Steven Vogt. They use very stable spectrographs to measure the tiny variations in a star's radial velocity that indicate the presence of an unseen companion. Other teams have found planets by searching for a partial eclipse of a star by a dark object. Now, with adaptive optics installed on large telescopes, astronomers hope to be able to obtain direct images of extrasolar planets.

So far they have not been successful, mainly because of the glare of the parent star. A Caltech team, using adaptive optics at the Palomar 5-meter telescope in 2003, detected two double and four single point sources close to the star Vega. At best, they were able to set upper limits to the masses of these objects, if they are in fact companions of Vega. But they could not say they had found a planet. Many others are trying.

In October 2003 a team led by physicist Reinhard Genzel (Max Planck Institute for Extraterrestrial Physics) reported the discovery of powerful infrared flares within a few light-minutes of the supermassive black hole at the center of the Milky Way. (Its mass is estimated at 3.6 million Suns.) Hot gases falling into the accretion disk around the hole most likely cause the flares. The flares repeat every seventeen minutes, which suggests that the black hole rotates with a period of about thirty minutes. If confirmed, this would be the first determination of a black hole's spin. These observations were made with the NAOS adaptive optics system on the 8.2-meter Kueyen telescope of the VLT.

A new era in astronomy has barely begun, as large mirrors are coupled to increasingly complex instruments and adaptive optics systems. We can expect many more exciting discoveries in the near future.

THE ASTRONOMER'S MICROSCOPE

ONE OF THE MOST impressive images I saw during the *Apollo* 11 mission was Neil Armstrong's footprint in the Moon's dust. I wonder what it would take to look for it? His boot was about 30 cm long, and the Moon's distance is about 380,000 km. So from our perch on Earth, his print spans an angle of 160 *micro-*arcseconds. (For comparison, Jupiter has a diameter of about 50 arcseconds.) Could the Hubble Space Telescope show us his print? No, its diffraction limit of resolution in blue light is about two hundred times too large. We would need a perfect telescope mirror 480 meters in diameter to see this relic.

At the moment no telescope that size exists. But as it happens, there is an instrument on Earth that might possibly do the trick. It's called SUSI, an acronym for the Sydney University Stellar Interferometer. It's one of a breed of new telescopes that, while a bit shy on light-gathering power, could resolve not only Armstrong's footprint but the disks of distant stars as well. It is, in a sense, an astronomer's microscope.

Only a few giant stars, like Betelgeuse in Orion, have had their portraits taken by the Hubble Space Telescope. We know from other observations that some stars have bright or dark spots, in analogy to sunspots, that some stars flare up in brightness, and that others eject high-speed winds. We suspect that new planets may be forming in the dusty envelopes around young stars. As-tronomers would like to study these phenomena with direct images if only they could see them more clearly.

The surfaces of stars are not the only interesting targets that are too small to resolve with the largest existing telescopes. We would like to see what's going on at the centers of galaxies where massive black holes reside. We would also like to find life somewhere else in the universe, and that entails finding suitable extrasolar planets. One promising method is to search for the periodic wobble in the position of a star that the gravitational pull of a planet would cause. To do that would require a resolution of a few tens of micro-arcseconds.

So despite the construction of many 8- and 10-meter telescopes, optical astronomers still don't have the angular resolution they would need for some research.

Radio astronomers faced a similar problem back in the 1950s. They were just starting to map the sky at meter wavelengths. Their first task was to determine the positions of many discrete sources and measure their sizes. To resolve a source as big as the Sun, however, they would need an antenna 100 meters in diameter, something beyond most astronomers' wildest dreams.

So they turned to interferometry. By combining the signals from two or more small antennas, spaced far apart, they were able to attain the higher resolution they desired. The angular resolution of such an array is fixed by the longest baseline between two antennas, not by the size of each antenna. The Very Long Baseline Array, the largest of these instruments, links ten antennas between Hawaii and the Virgin Islands and can resolve *milli*-arcseconds.

Optical astronomers are now following the same path toward superresolution. The trail was blazed in 1970 by some critical experiments in France. In the 1980s several teams began to assemble arrays of small optical telescopes that were capable of extraordinary performance. Then in the 1990s telescopes like the Keck I and II or the four units of the VLT were hooked together as interferometers. Lately, smaller telescopes are being added to these giants in hybrid arrays. With these new tools astronomers are making some interesting discoveries.

To appreciate these developments we need to step back a few paces, to understand the interference of light and how interferometers work. We'll spend some time with the radio astronomers to recall some of their important insights. Then we can see how the optical astronomers are progressing and survey some of their scientific results.

THOMAS YOUNG

Some of us may have noticed how an ocean wave, approaching parallel to the shore, will spread out in all directions when it passes through the narrow entrance to a harbor. This is a homely example of diffraction (fig. 10.1). When the diffracted waves arrive at the beach, they are reflected back toward the harbor entrance, where they overlap (or *interfere*) with waves that are just arriving. When the peaks of two waves overlap, we see them add together in a high splash. Where a peak and a trough overlap, the result is a flat sea.

Physicists of the eighteenth century were well acquainted with such examples of the diffraction and interference of ocean waves. They had also seen similar effects among sound waves but not among light waves. In fact, they were still debating the physical nature of light.

Sir Isaac Newton had proposed, a century earlier, that light consists of myriad tiny particles that he called *corpuscles*. With this hypothesis he could explain the reflection and refraction of light and why light travels in straight lines. Christian Huygens, an eminent Dutch scientist, proposed instead a wave theory of light. He too could explain reflection and refraction but had no idea what was waving. Newton was dubious about Huygens's theory because diffraction, the bending of waves by an obstacle, hadn't been observed in light.

Thomas Young, an English scientist, devised a simple experiment that

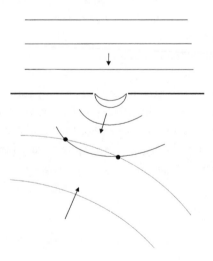

Fig. 10.1. A train of ocean waves approaches the narrow mouth of a harbor *(top)*. When a wave squeezes through the entrance, it spreads sideways by the phenomenon of diffraction. Eventually these waves reach the shore, where they are reflected back toward the entrance *(thin lines)*. When they meet incoming waves, they overlap. Interference between waves causes them to add or cancel.

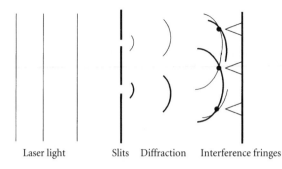

Fig. 10.2. In a modern version of Young's interference experiment, monochromatic light from a laser passes through two narrow slits. The light waves are first diffracted and then interfere to form a pattern of fringes on the far screen. The width of a fringe is a measure of the separation of the slits.

Laser light   Slits   Diffraction   Interference fringes

demonstrated both diffraction and interference of light and helped to establish the wave theory. Young had been a child prodigy who could read English at the age of two and Latin at six. At twenty-eight he was already a professor at the Royal Institution, lecturing on everything from astronomy to zoology. He studied medicine and practiced as a physician, but his true love was physics. After inheriting a fortune he was able to devote himself entirely to physics. He carried out numerous experiments on light and sound in his private laboratory. Later in his career he became an accomplished Egyptologist who interpreted some of the key hieroglyphics of the recently discovered Rosetta Stone.

Around 1800 Young constructed his elegant double-slit experiment (fig. 10.2). Monochromatic light from a distant source falls on two very narrow slits in a screen. A pattern of light and dark fringes is seen on a second screen, which would not be expected if light were composed of Newton's corpuscles. Young explained his result in terms of the wave theory as follows.

A light wave from the source reaches both slits with the same phase. The segments of the wave that pass through the slits are diffracted. That is, they tend to expand spherically, as though each slit was itself a source of light. As a result, the waves from the slits spread over the second screen and overlap. The path lengths from the slits to the center of the second screen are equal, and so the waves from the slits arrive in phase. Therefore they add together to produce a bright fringe. This is called *constructive interference.* It occurs wherever the path lengths from the slits to a point on the second screen differ by a whole number of wavelengths. At a place on the screen where the path lengths differ by an odd number of half wavelengths, one wave's trough overlaps the other wave's peak and they cancel each other in *destructive interference.* That is the origin of the dark fringes.

Thomas learned that the width of a fringe depends on the ratio of the wavelength w and the separation S of the slits, or w/S. The larger the separation, the narrower the width. On the other hand, the width of the whole pattern of fringes increases as the slits are narrowed.

Thomas presented his results in a lecture to the Royal Society of London in 1803, but they made surprisingly little impression on the assembled dignitaries. Newton's prestige was too overwhelming. Only after August Jean Fresnel, a French physicist, reported his experiments with polarized light and published his mathematical theory of diffraction and interference was the wave hypothesis fully accepted.

THE STELLAR INTERFEROMETER

Armand Hippolyte Louis Fizeau, another great French physicist, suggested in 1868 that stellar diameters might be measured with a version of Young's two-slit experiment. Edouard Stephan followed up on this suggestion by placing a mask with two holes over the 0.8-meter mirror of a telescope. He saw no fringes and concluded, correctly, that stellar diameters must be smaller than one-tenth of an arcsecond.

Albert Michelson, an American physicist, was the next to try. Light was the passion of his life. He had previously made the most accurate measurement of the speed of light, and with the chemist Edward Morley he had demonstrated that the *ether*, the medium in which light was supposed to travel, does not exist. He was strangely pessimistic about the future of physics, however, despite his own seminal discovery. In 1894 he stated, "The more important fundamental laws and facts of physical science have all been discovered. . . . Our future discoveries must be looked for in the sixth place of decimals." He would soon contradict his own prediction.

In 1921 Michelson mounted a 6-meter steel beam at the front end of Mount Wilson's 100-inch (2.5-meter) telescope. Light from a star was relayed in two paths to the main telescope mirror by four small mirrors. Michelson varied the path length of one beam until he could see fringes at the telescope's focus.

A true point source would produce fringes of high contrast, as in Young's experiment. A resolvable star, however, would produce fringes of lower contrast (or *visibility*), because fringes from different parts of the stellar disk would overlap out of phase. By varying the path length of one beam, Michelson could search for minimum visibility. The angular diameter of the star was then equal (in radians) to the ratio of the wavelength and this critical path length difference. In this way, he and his colleague Francis Pease measured the diameter of Betelgeuse (the brightest star in Orion) as 0.047 arcsecond.

This was at least ten times smaller than the 100-inch alone could resolve in even the best seeing. The power of the interferometer lies in the long separation (or *baseline*) between its reflecting elements. The longer the baseline, the narrower the angular spacing between fringes, and therefore the higher the resolution.

Michelson and Pease went on to measure the diameters of six other stars. And then progress stalled. Pease attempted to extend their results with an independently mounted 15-meter steel beam but was frustrated by a variety of factors. The beam sagged and vibrated, and the seeing blurred his fringes. He abandoned his efforts without obtaining any significant results.

Further progress in optical interferometry would have to wait for improvements in technology. But beginning in 1945, radio astronomers began to use interferometers out of necessity. They learned several techniques that optical astronomers would adopt: nonredundant baselines, aperture synthesis, delay tracking, and phase closure.

BEGINNINGS IN BRITAIN

As we learned in chapter 3, radio astronomy took off after World War II, when radar physicists applied their newly won skills to astronomical problems. After the first few discoveries at meter wavelengths by Grote Reber and Karl Jansky, several groups, particularly in Australia and Britain, settled down to find and catalog the thousands of discrete sources in the sky.

They faced two challenging technical problems. To detect the faint emissions, they needed large collecting areas; to fix the source positions precisely, so that they might identify them with optical objects, they needed high angular resolution. A telescope that satisfied both requirements would cost more than most astronomers could afford. So each group chose to emphasize one or the other goal.

Bernard Lovell, at the University of Manchester, chose to search for faint sources with a large antenna. He first built a fixed wire bowl 66 meters in diameter that could only look straight up. Despite its limited view, it was able to

detect the radio emission of a dozen sources, including the Andromeda Nebula. Lovell began to campaign for a large steerable dish, and by 1957 he had
completed the Mark I, a 76-meter steerable dish at Jodrell Bank, near Manchester, England. The Mark I eventually cataloged thousands of radio sources
and played an important role in the discovery of quasars.

At Cambridge University, Martin Ryle wasn't about to wait for a large telescope. He decided to pin down the brightest sources first, with an interferometer. He would sacrifice collecting area for higher angular resolution. So in the
late 1940s he built a radio version of a Michelson interferometer (*Monthly Notices of the Royal Astronomical Society* 110, 508, 1950). It had two fixed antennas
separated by 400 meters on an east-west baseline. A radio receiver tuned
sharply to 3.7 meters detected the signal, which was recorded on a chart recorder.

Such an arrangement is sensitive to cosmic radio waves only in a narrow
fan-shaped beam that is analogous to the fringe pattern of Young's two-slit experiment (fig. 10.3). In effect the interference fringes are projected on the sky,
and represent angles at which constructive and destructive interference occurs
between the signals from the two antennas.

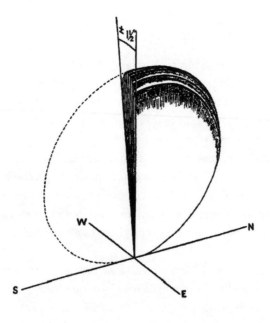

Fig. 10.3. Ryle's two-antenna
interferometer had a fan-
shaped beam with narrow
fringes that are analogous to
Young's interference fringes.

Ryle's beam was narrow in the east-west direction and pointed due south at the long vertical strip on the sky called the *meridian*. As the Earth rotated, discrete sources would sweep through the fringes of the beam. If a source was smaller than the half-degree separation of the fringes, he would see a sinusoidal chart recording. Noting the time when the fringe amplitude was a maximum, Ryle could determine the source's position in the east-west direction, and from the period of the sinusoidal tracing he could estimate its north-south location as well (note 10.1).

In 1948 Ryle and his colleague Francis Graham Smith discovered the brightest radio source in the sky, Cassiopeia A, with this instrument. By 1950 the Cambridge team had located fifty radio sources over the Northern Hemisphere at a wavelength of 3.7 meters. They had gotten off to a fast start.

A two-antenna interferometer with an east-west baseline measures the brightness of a source only in north-south strips. Ryle wanted more. So in 1955 he and Antony Hewish built an interferometer with four fixed antennas at the corners of a rectangle measuring 580 by 49 meters. It had narrow beams in both the east-west and north-south directions, and by combining the signals they could obtain both the positions and two dimensions of a source. This instrument found over four hundred sources that were published in the famous 3C catalog.

Later on Ryle learned to track a source across the sky, even though his simple dipole antennas could not be steered. Figure 10.4 shows how he did that. Suppose he wanted to aim his beam at a source that lies west of the meridian. He had to ensure that the signals from both antennas were in phase, otherwise they would not interfere constructively. But a wave had to travel farther to reach the east antenna than the west antenna, and during this travel time its sinusoidal phase changed. So Ryle introduced an electronic time delay in the west arm of the interferometer that compensated for the extra travel time. That effectively points the beam at the source, and by varying the delay, the source could be tracked across the sky. Delay tracking would be adopted by optical astronomers, as we shall see.

THE ROAD TO APERTURE SYNTHESIS

Australian physicists and engineers were also pioneering in radio astronomy just after the war. Perhaps their most important contribution in these early

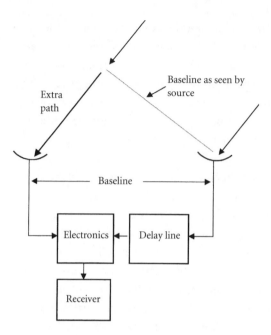

Fig. 10.4. A radio wave arriving at an angle reaches one antenna of an interferometer before the other. To obtain interference of the two signals, however, the path lengths must be equal. So a variable time delay is introduced into the arm from the antenna on the right to compensate for the longer time the wave needs to arrive at the left antenna. The magnitude of the delay is a measure of the phase of the radio wave.

days was not a discovery but an idea. In 1947 Lewis McCready, Joseph Pawsey, and Ruby Payne-Scott introduced a concept that would dominate radio astronomy and later, optical interferometry. It is based on the work conducted over a century earlier by one of the greatest French mathematicians.

Jean Baptiste Joseph Fourier proved that the shape of any arbitrary object, like a person's face or a mountain range, could be represented by a sum of sine and cosine functions of different spatial wavelengths and amplitudes. The long-wavelength waves overlap to outline the broad features, and the short wavelengths combine to shape the finer details (fig. 10.5).

From this viewpoint, one can think of the brightness distribution of a broad radio source as the sum of overlapping sine and cosine waves. McCready, Pawsey, and Payne-Scott pointed out that a two-antenna interferometer responds to just one spatial wavelength in the source. The length of the baseline determines this unique spatial wavelength. So if one could measure the source brightness at many different baselines and also at many different orientations of a baseline, one could find all the Fourier sine and cosine waves that determine the shape of the source. Then one could add up these components in a computer to reconstruct a two-dimensional *image* of the source.

Patrick O'Brien, a young colleague of Ryle, was one of the first to apply this idea. In 1951 he set up a two-antenna interferometer on an east-west line to observe the Sun around noon. Each day during the summer he changed the separation of the antennas. Then he shifted one antenna off the east-west line to change the orientation of the baseline and repeated the whole sequence of observations. In this way he carried out the program the Australians had suggested.

He processed the data on the EDSAC, an early model of an electronic computer. After fifteen hours the computer spat out profiles of the Sun's brightness at different angles across its disk. O'Brien was able to confirm that, near sunspot minimum, the Sun's atmosphere extends further in the equatorial plane than at the poles, and the extension increases at longer radio wavelengths.

Martin Ryle had suggested O'Brien's experiment, and its success convinced him of the potential of Fourier methods. His immediate problem, however, was a shortage of collecting area. To catalog fainter sources, he would need more antenna area. A large parabola, like Lovell's at Jodrell Bank, was a possible solution, if he could afford it. But observing at a wavelength of a few me-

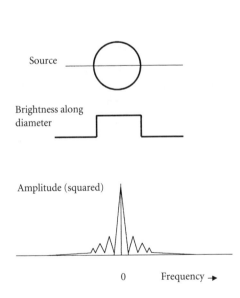

Source

Brightness along diameter

Amplitude (squared)

0   Frequency →

Fig. 10.5. As an example of Fourier analysis, consider a circular radio source *(top panel)*. The brightness along the diameter of the source is shown in the middle panel. The Fourier representation of this uniform brightness distribution is a sum of cosine functions of different spatial frequencies. In the bottom panel we see the amplitude (squared) of the cosine functions as a function of frequency. The low frequencies (near zero) define the gross size of the source, while the high frequencies, which are represented with lower amplitudes, define the sharp edge of the source. In radio interferometry, a measurement at each baseline yields the amplitude and phase of a particular spatial frequency. Then by combining the Fourier information in a computer one can reconstruct the shape of the source.

ters, where the sources were strongest, he would need a gigantic dish to obtain an angular resolution of a few arcminutes, and that was out of the question. An alternative solution was to build a large interferometer, as the Australians were doing.

Bernard Mills, Alexander Shain, and William Christiansen had each built a huge interferometer in the shape of a cross, with north-south and east-west arms hundreds of meters long. Mills and his colleagues combined the beams from the two arms to form a "pencil beam" about 50 arcminutes in width. They scanned the sky, using the pencil beam like a searchlight, and determined the positions of over two thousand radio sources. It was slow, laborious work. A pencil beam was not well-suited for surveys.

In 1960 Ryle and Antony Hewish proposed a much more efficient technique. In principle it could offer much more collecting area than, say, Mills's Cross, while keeping the cross's high angular resolution.

APERTURE SYNTHESIS

Mills's Cross is bounded by a rectangle hundreds of meters on a side. The expensive way to increase the collecting area of his array would be to fill in this rectangle with small antennas. Suppose instead that one starts anew. Suppose that one places a small antenna, A, somewhere in a rectangle and moves a similar small antenna, B, to another position in the rectangle. At each position of B, the pair can be used as a two-element interferometer, with a specific baseline length and orientation, to measure the amplitude and phase of a source as it rotates through the fan-shaped beam.

One can repeat this procedure and make similar measurements at all the baselines and orientations that the rectangle contains. Then, as O'Brien demonstrated, one can use Fourier methods to calculate the brightness distribution of the source.

The beauty of this method is that, after the calculation, one obtains the same resolution and *almost the same signal strength* as with a fully filled antenna the size of the rectangle. In effect, the procedure of Ryle and Hewish *synthesizes* a large collecting area, or *aperture*.

What's the catch? Wouldn't such a series of measurements at different positions take too long to be practical? Ryle and Hewish pointed out, in response, that at each arrangement of the two antennas, A and B, a wide area of the sky could be scanned by steering the fan-shaped beam in different directions (note 10.2). To accomplish the same task, at the same angular resolution, a fully filled aperture would have to step its pencil beam sequentially over the same area of the sky. Ryle and Hewish demonstrated that the total observing time required to scan the area with a synthesized aperture was only twice as long as with a fully filled aperture.

Similarly, the effective collecting area of the synthesized aperture is smaller, but not much smaller than that of the corresponding filled aperture. If the filled aperture is a square of side D and the movable antenna has a side d, the synthesized area is smaller by the ratio $3d/D$, not, as one might expect, d^2/D^2.

WHY BOTHER TO MOVE AT ALL?

Ryle soon discovered that there was no need to shuffle antennas around like railroad cars to synthesize a large aperture. The rotation of the Earth would do it for him.

Almost all previous interferometers had used the Earth's rotation simply to bring sources into a fixed beam. Ryle realized that the projected length of a baseline, as seen from the source, varies as the Earth rotates (fig. 10.6). In addition, if one could track the source, as it crosses from east to west, the fringes of one's interferometer would *rotate* over the source by 180 degrees every twelve hours. The combination of these two effects would allow one to sample many spatial wavelengths, at many angles on the sky (to see how many, read note 10.3). That would yield a great deal of information on the Fourier repre-

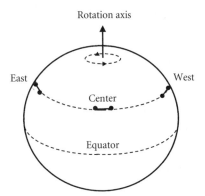

Fig. 10.6. Imagine that you are a radio source looking back at Earth. From your point of view, the baseline of an interferometer varies in length as the Earth rotates. It is this variation that allows the observer on Earth to sample different spatial wavelengths in the source's two-dimensional brightness distribution.

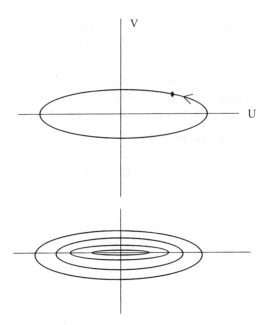

Fig. 10.7. *Top:* A two-element interferometer with a specific baseline and orientation samples two spatial frequencies (*u* and *v*) in the brightness distribution of a broad source: *u* in the x direction on the sky and *v* in the y direction. As the Earth rotates, the projected baseline and its orientation change, and so the values of *u* and *v* trace an ellipse in the *u,v* plane. *Bottom:* An array with many baselines collects a nested set of ellipses. These pairs of frequencies can be used in a partial Fourier reconstruction of the source distribution.

sentation of the source (fig. 10.7). Then, using a computer, one could reconstruct a two-dimensional map of the brightness of the source. (This technique is an extension of what William Christiansen and John Warburton had done for the Sun back in 1953. See note 10.4.)

This, then, was the principle of aperture synthesis. The Cambridge group, under Ryle's direction, went on to make many important discoveries with a series of increasingly powerful interferometers, using this principle. In 1974 Ryle shared the Nobel Prize in physics with Hewish for their contributions to radio astronomy.

INTERFEROMETERS AROUND THE WORLD

Earth-rotation aperture synthesis introduced a revolution in the way radio astronomers observed the heavens. Beginning in the 1960s, they built larger and larger interferometric arrays to take advantage of the principle. Instead of merely two antennas, these interferometers consisted of many identical units, arranged in such a way as to obtain many Fourier components simultaneously. The Dutch built a fourteen-dish array at Westerbork, the Americans built the

VLA, the Very Large Array (a twenty-seven-dish array in the shape of a Y), the Canadians a seven-dish, the Australians a six-dish—the list goes on and on. Figure 3.4 offers an example of the kind of detail a large array can reveal in a distant radio source.

In a multielement array, each combination of two elements constitutes a specific baseline. So adding one more element to an N-element array adds N new baselines, and the number of baselines grows as $N(N-1)/2$. For example, 3 antennas in a triangle can be paired to form 3 baselines, 7 antennas in a polygon form 21, and 27 antennas (as in the VLA) form 351. For maximum efficiency, none of the baselines should have the same length or the same orientation; they should be *nonredundant*. In any array of discrete elements there are gaps in the coverage of baseline lengths, but the rotation of the Earth helps to fill these in (see note 10.3).

WHY NOT US?

Optical astronomers must have watched with envy as the radio astronomers achieved ever-higher angular resolution with large arrays and with aperture synthesis. Why couldn't the same methods be used at optical wavelengths? After all, Michelson and Pease had shown the way.

Extending their methods had proved to be extremely difficult, however. The reason lies in the enormous ratio of the wavelengths of radio waves and visible light, a factor of a million. That simple fact erects a variety of technical hurdles. The path lengths in an optical interferometer must be equal to within a wavelength of light. So the mechanical dimensions must be stable to within one-twentieth of the width of a human hair. Vibration, thermal expansion, bad seeing, and misalignments are much more serious sources of error.

Even worse, the frequency of light (typically 300,000 GHz) is much higher than the best electronics can handle. That means that light, unlike radio waves, cannot be amplified while preserving its phase. Instead, the incident beams of starlight must be brought together optically to enable them to interfere. The whole enterprise becomes extremely daunting.

With the passage of time, technology improved on all fronts, and another attempt at long-baseline optical interferometry appeared to be feasible. In the early 1970s Antoine Labeyrie, the French scientist whom we have met before, began experimenting at the Meudon Observatory, near Paris. From the beginning he had in mind building arrays of small telescopes analogous to the radio interferometers and employing aperture synthesis.

In 1973 he built a two-mirror interferometer with a fixed north-south baseline 12 meters long, near the Nice Observatory. In principle his instrument was a Michelson stellar interferometer, but it had several essential innovations. His 25-cm mirrors rotated in alt-azimuth mounts, under computer control, to track a star. The two beams were directed to a central laboratory where they were combined with small optics to produce Young's fringes. A state-of-the-art photon-counting television camera was used to record the fringe contrast.

To see fringes at high contrast, Labeyrie had to adjust the path length of the light in each arm of the interferometer to be exactly equal. But as the mirrors tracked a star, the difference in path lengths changed continuously. Radio astronomers solved this problem by inserting a variable electronic delay in the shorter arm. Labeyrie had no such option, so he introduced an optical delay. He mounted his combination optics on a small tray that slid smoothly along a linear track. As these optics moved, one arm shortened while the other lengthened. This was the first use of an *optical delay line*, a key feature in the success of his experiment.

In 1975 Labeyrie obtained high-contrast fringes of the star Vega, which demonstrated a resolution of 5 milli-arcseconds. Two years later he built an improved version of this instrument at the CERGA site, near Grasse, France. Here the 25-cm mirrors were mounted on precision-ground rods to provide for a variable baseline as long as 20 meters. By searching for the baseline at which the visibility of the fringes was a minimum, Labeyrie and his colleagues were able to determine the angular diameters of Capella A and B, a very close binary system, with a precision of 1 to 2 milli-arcseconds. Moreover, they used a crude version of aperture synthesis to verify the calculated orbital parameters of the binary system. The two stars are separated by a mere 50 milli-arcseconds.

With these successes, Labeyrie immediately began to build Project Argus, an optical array consisting of four 1.5-meter mirrors on crossed variable baselines. To economize on the telescope mounts he built an experimental spherical bubble of ferro-concrete 3.5 meters in diameter and precision-ground. This *boule* sat in a horizontal ring whose rollers kept the mirror pointed at a star, under computer control. A pair of such telescopes (named G12T) were completed and produced some high-resolution observations of stellar diameters and atmospheres.

Project Argus was never finished, but Labeyrie continued to spin out radically new ideas. By 1990 he was proposing to build an Optical Very Large Array of twenty-seven telescopes and a Moon-based system as well.

AN IDEA AHEAD OF ITS TIME

Labeyrie's pioneering efforts ignited enormous interest among astronomers. Everyone began to recognize that arrays of optical telescopes could work, with the potential for huge advances in angular resolution. And with the advent of lasers and faster computers, the technical problems would be easier to solve. The European Southern Observatory and the Keck consortium both planned to couple their big telescopes as interferometers. Several groups, in Britain, Australia, and the United States, began designing optical arrays of small telescopes. But the Multi-Mirror Telescope (MMT) Observatory was the first to put a plan into action.

Its success was primarily the work of Jacques Beckers, a Dutch-born astronomer with a talent for creative instrumentation. After a fifteen-year career in solar physics, Beckers shifted his attention to the challenging field of optical interferometry. In 1979 he assumed the directorship of the MMT Observatory.

As we learned in chapter 4, the MMT was one of the first so-called advanced-technology telescopes. It consisted of six 1.8-meter mirrors on a common alt-azimuth mount (see fig. 4.10). The light from all six was combined to give the MMT the effective collecting area of a 4.4-meter mirror, but with the angular resolution of a 1.8-meter.

When Beckers arrived he set the goal of converting the telescope to an optical interferometer that would achieve the resolution of a 7-meter baseline over a small field of view. An even more ambitious goal was to make the six mirrors perform as parts of a single mirror 4.4 meters in diameter, with a wide field of view. That would require phasing the individual mirrors so that their path lengths to any point on the focal plane would be exactly equal.

Beckers and his team began by linking two mirrors 5.1 meters apart as a Michelson interferometer. Beckers experimented endlessly with several methods for equalizing their path lengths. Gradually, several more mirrors were added to the array. After five years of hard work, Beckers and Keith Hege were able to combine all six mirrors as an interferometric array. The telescope now had the diffraction-limited resolution of a 7-meter over a wide field of view. Even better, they were able to make the MMT perform as a single mirror 4 meters in diameter by tilting the focal plane. They exhibited diffraction-limited images of such stars as Alpha and Gamma Orionis, and the famous binary system of Capella.

The MMT served as a valuable test bed in attacking the dozens of problems involved in making an array of meter-class telescopes perform as an interferometer. But by the mid-1980s Roger Angel's spin-casting experiments were showing the way toward much larger monolithic mirrors. Why bother with a finicky cluster of small mirrors if one could acquire a single large one? And with the further development of adaptive optics, a large mirror could reach diffraction-limited resolution. The MMT's board of directors decided to terminate the interferometry program and to convert the MMT to a 6-meter monolith. Beckers disagreed vehemently with this decision and resigned from the MMT in 1985. He carried his fund of experience to another large project, the U.S. National New Technology Telescope (NNTT).

THE NATIONAL NEW TECHNOLOGY TELESCOPE

The National Optical Astronomy Observatory (NOAO) in Tucson serves the needs of a huge population of astronomers by providing first-class telescopes and equipment. In the early 1980s the largest telescopes in its inventory were the 4-meter Blanco in Chile and the 4-meter Mayall on Kitt Peak. The staff was now preparing to build the next large telescope. Two designs were being considered: a cluster of telescopes similar to the MMT, and a large segmented mirror similar to that proposed for the Keck Observatory.

After many debates, a cluster of four 8-meter mirrors on a common mount was chosen, similar to the MMT. It would have the collecting area of a 16-meter mirror, and acting as an interferometer it would have a maximum baseline of 22.1 meters. At a wavelength of 2 microns, its angular resolution would reach 20 milli-arcseconds. With the success of the MMT as a precedent and with Beckers on the staff, its prospects seemed excellent.

In time, however, the board of directors of the NOAO became concerned about the telescope's technical feasibility, cost, and schedule. A decision was taken to abandon the MMT-type design and plan instead for two independent 8-meter telescopes, the Gemini project that we described in chapter 8. After three years of work on the NNTT, Beckers left the NOAO in 1988 and moved once again to another promising project, the Very Large Telescope Interferometer (VLTI) of the European Southern Observatory (ESO).

AN ACRE OF MIRRORS?

The idea of using giant mirrors on a common mount for interferometry has survived in the Large Binocular Telescope, which we described in chapter 7. However, most astronomers working on optical interferometry have followed Labeyrie's example and built arrays of small telescopes on very long baselines.

All these arrays have common advantages and constraints. Their overwhelming advantage, of course, is their extraordinary angular resolution. Some of these arrays are capable of resolving a fraction of a milli-arcsecond, which allows imaging the surface of a star.

The biggest constraint so far is bad seeing, that old enemy. If the fringes twinkle, shift, or fade, a precise measurement is impossible. To avoid seeing effects, most interferometers employ mirrors not much larger than the size of a typical air cell (or Fried's parameter r_0), say a few tens of centimeters. Therefore, only stars brighter than magnitude 10 are accessible. Meter-sized mirrors would be usable if and when adaptive optics were installed, but that is an expensive solution.

Larger mirrors have another downside, unfortunately. The field of view of an array is limited by the diffraction limit of its smallest element. An array with 1-meter mirrors would be limited to a field of 0.2 arcsecond at a wavelength of 1 micron. That's certainly large enough for some interesting science, but not for everything one might think of.

MORE ACTION AT CAMBRIDGE

It should be no surprise that the first optical array to employ aperture synthesis was built at Cambridge University, where Ryle and Hewish had invented the method. John Baldwin and his colleagues began to plan the Cambridge Optical Aperture Synthesis Telescope (COAST) in 1986. Construction began in 1988 and was completed in three years at a cost of £850,000.

COAST consisted originally of four 50-cm siderostats arranged in a Y that fed fixed horizontal telescopes. (A *siderostat* is essentially a flat mirror, tilted at the proper angle for the star and able to rotate to track it.) The siderostats are movable along rails, and the maximum baseline was initially 12 meters. The beams from the 40-cm Cassegrain telescopes shine down a tunnel, in air, to an underground lab, where optical delay lines equalize the path lengths to a common beam-combining set of optics. A fast readout detector records the fringes, and the data are stored immediately in a computer. The array works at infrared wavelengths of 0.8 and 2.2 microns.

COAST produced its first fringes of a star in 1991, and in 1993 Baldwin and his team were the first to demonstrate *phase closure,* a method well known to radio astronomers. By combining the phases from three or more telescopes in a polygon, the team was able to cancel the phase error introduced at each telescope by the atmosphere (note 10.5). That was a big step toward achieving higher resolution.

The COAST team has the typical scientific aims of most groups in this field. They want to study activity on the surfaces of stars, the gaseous envelopes of stars, stellar pulsations, and the interactions in multiple-star systems.

COAST passed an important milestone in 1995 when it obtained images of the binary Capella by aperture synthesis and phase closure (fig. 10.8). Not only were the disks of the stars resolved, but their orbital motion was followed over fifteen days. A resolution of 20 milli-arcseconds was achieved with a maximum baseline of 6 meters.

In 1997 the team obtained images of Betelgeuse, one of the favorite targets of such experimenters. For the first time an interferometer detected the variation of brightness over the stellar disk, the *limb-darkening.* Then, in 2000, bright areas on the disk of the star were resolved in three colors in the red and infrared bands (fig. 10.9). These regions are apparently connected with convection of the stellar interior.

CH Cygni is a *symbiotic* star. It is a red giant star that is continually spewing gas off to a close companion. The COAST has revealed that the red giant has an elliptical shape, due perhaps to a high rate of rotation.

A fifth telescope was added to the array in June 1998, and baselines were extended, first to 48 meters, then in 2002 to 67 meters. Most recently the COAST team has joined with teams based in New Mexico and Puerto Rico and with the Naval Observatory in Washington, D.C., to form the Magdalena Ridge Consortium. They plan to build a $40 million array of eight to ten telescopes, 1.4 meters in diameter, on baselines as long as 400 meters. At a wavelength of 2.2 microns, this array could resolve one-tenth of a milli-arcsecond, three hundred times smaller than the limit of the Hubble Space Telescope. The consortium plans to hunt for extraterrestrial planets by searching for the wobble in the position of stars.

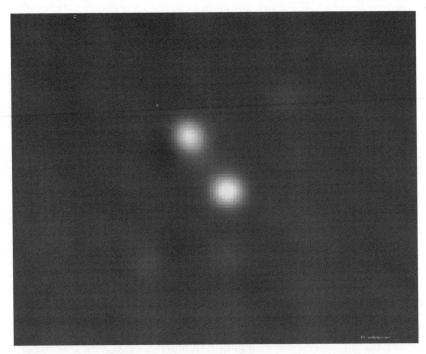

Fig. 10.8. The first aperture synthesis image of a close binary, Capella, obtained with the interferometer of the Cambridge Optical Aperture Synthesis Telescope. The separation of the stars is about 50 milli-arcseconds.

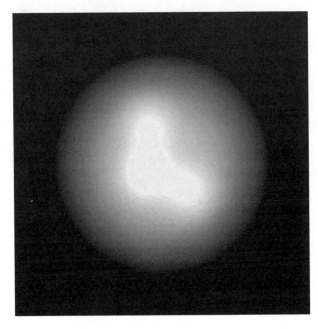

Fig. 10.9. Bright spots on the surface of Betelgeuse, at a wavelength of 700 nm, obtained with the COAST interferometer

THE AUSTRALIAN ENTRANT

Sydney University has a long tradition in optical interferometry. Back in the 1960s and 1970s, Hanbury Brown and his associates were measuring the diameters of blue stars with a novel technique called *intensity interferometry*. The scheme had several advantages but was limited to stars brighter than magnitude 2. In the early 1980s the staff, led by John Davis, began to plan a simple two-telescope interferometer that could be more flexible. It would have the longest baseline of any contemplated at that time (or since), 640 meters, and would have the whole Southern Hemisphere to itself. Funds were granted in 1985, construction began in 1987, and the first fringes were obtained in 1992.

The Sydney University Stellar Interferometer (SUSI) is located 20 km from Narrabri, in New South Wales. Narrabri is the site of the Australia Telescope, a major radio interferometer, and the university team has benefited from close ties to the radio astronomers.

SUSI employs eleven siderostats at fixed positions on a north-south baseline. Only two of these are used during an observation, but the observer can choose baselines ranging from 5 to 640 meters. The starlight is reflected down an evacuated pipe to a central laboratory where the beams are combined.

To minimize the effects of seeing on the fringes, the team chose to use very small telescope mirrors, only 14 cm in effective aperture. In addition, a tip-tilt mirror is provided with each telescope to stabilize the beam. But at a wavelength of 0.45 micron (which is quite short for this kind of work), SUSI can achieve a range of resolutions from 20 milli-arcseconds to 70 *micro*-arcseconds. The downside is SUSI's sensitivity. It is limited to stars of magnitude 2 at blue wavelengths and magnitude 5 in the near infrared.

Finding a star with such an apparatus can be something of a challenge. The field of view of the interferometer is only 3 arcseconds, one-twentieth of the maximum diameter of Venus, so a separate optical system is needed. It steals a small fraction of the light from one telescope to feed a sensitive CCD camera. With that arrangement, an observer can see a field of 8 arcminutes, about one-quarter of the diameter of the Moon.

In principle SUSI could produce images of stellar surfaces by aperture synthesis, but as of 2001 the team hadn't attempted that. Instead they are focusing on determining the orbital parameters and masses of well-known spectroscopic binary systems. In a spectroscopic binary the stars are too close for ordinary telescopes to resolve, but their binary nature is revealed by the periodic oscillations of their overlapping spectra. By combining spectroscopic periods and velocities with interferometric measurements of the stars' separation, the team can determine the mass of each star. In July 2003 John Davis reported results on the double-lined spectroscopic binary Beta Centauri. The bright blue giant star's mass could be determined with a precision of a few percentage points.

THE CENTER FOR HIGH ANGULAR RESOLUTION ASTRONOMY

Hal McAllister has also been in this business for a long time. Beginning in 1977 and continuing until 1998, this professor of physics and his team at Georgia State University used speckle interferometry to resolve close stellar binary systems. They observed wherever telescope time was available and eventually gained access to the 4-meter-class telescopes in Arizona, California, Chile, and Hawaii. They determined the masses, orbits, and periods of thousands of binary systems, a body of data that is immensely valuable for the study of star formation.

This program gave McAllister the credibility he needed to reach a more ambitious goal. In 1984 he established the Center for High Angular Resolution Astronomy (CHARA) as a base for the promotion, construction, and operation of a large interferometer. It was a long struggle. He began planning the array in 1985, but it took a decade of proposals and reviews to acquire the necessary funds. Finally in 1996 the National Science Foundation granted over $6 million for a five-telescope array, and the university matched the funds. Mount Wilson Observatory was chosen as the site, and ground was broken in 1996. Supplements granted in 1998 by the Keck Foundation and the Packard Foundation allowed McAllister to add a sixth telescope.

The CHARA array now consists of six 1-meter telescopes arranged in a Y. The telescopes can't be moved, but they stake out fifteen nonredundant baselines ranging from 34 to 354 meters. At visible wavelengths the array has a maximum resolution of 280 micro-arcseconds, the angle a dime makes at a distance of 7,000 km.

Like all such arrays, the interferometer is a complex system with three main components: the telescopes, the delay lines, and the combination optics. Each

Fig. 10.10. The optical delay lines of the Center for High Angular Resolution Astronomy array. The optics in the foreground are fixed. In the background you can see the carts that slide along the tracks.

Cassegrain telescope has a tip-tilting secondary mirror and an alt-azimuth mount, both under computer control. The six beams travel through evacuated pipes to a Beam Synthesis Facility, a state-of-the-art laboratory 100 meters long. The beams then exit the vacuum and enter the delay line tunnel, where the path lengths from the telescopes to the combination optics are equalized.

CHARA has six delay lines, each with a moving cart that contains the reflective optics (fig. 10.10). As the cart slides smoothly along its rails, the path length of a beam gradually increases. A fringe-tracking mechanism feeds information back to the carts to ensure that they move at an absolutely constant rate. That allows observations to be continued for several hours. Additional systems align the beams and change their focal ratios before they enter the combination optics. They resemble a railroad switchyard, with separate sixfold paths for visible and infrared wavelengths.

The CHARA array has an unusual system designed to use as much of the light as possible. For beams to interfere and form fringes, their light must be nearly monochromatic. Ordinarily that meant discarding all but a narrow band of wavelengths. In this array fringes of many different colors are separated by a spectrograph and recorded independently. This feature allows the array to detect fainter stars. Even so, the limiting magnitude is only about 10.

With optical paths hundreds of meters long, the equipment is extremely sensitive to vibration, either manmade or natural, and extraordinary measures are necessary to isolate the optics. So the Beam Synthesis Facility was designed as a "building within a building," thermally insulated and cushioned against vibration.

The CHARA array produced fringes of Sirius at its longest baseline for the first time in September 2001 and, after a year of fine-tuning, began science operations in December 2002. It will be used to determine stellar diameters, luminosities, masses, and temperatures. In addition, the team hopes to calibrate the stellar distance scale by observing the expansion of nova shells and the pulsation of Cepheid variable stars. Observing the surface and circumstellar regions of stars is another important goal.

In 2003 McAllister described the start of a major program on spectroscopic binary systems. A catalog of a few thousand known binaries has been compiled in preparation for observations to determine masses, radii, and temperatures.

In January 2005 the CHARA array scored another first. Regulus, the brightest star in the constellation Leo, completes a rotation in less than one day. Its equator spins so rapidly that the centrifugal force at the surface nearly cancels the gravitational force. Theorists predicted that, as a result, Regulus's equator must be cooler (and therefore darker) than its poles. Using the CHARA array, astronomers have now confirmed that the star's poles are five times brighter than the equator. Its measured equatorial diameter is also one-third larger than the polar diameter. This result is a striking illustration of the power of the new array.

HAIL TO THE NAVY

In the past decade, U.S. Naval Observatory, the Naval Research Laboratory, and Lowell Observatory have built two elaborate optical interferometers on a mesa near Flagstaff, Arizona. One array is intended for *astrometry,* the measurement of the positions and angular motions of stars. The other array is designed to obtain images of stellar surfaces.

You might wonder why the navy is so heavily engaged in interferometry.

One answer is that the navy has always searched for ways to improve celestial navigation. Today, the Global Positioning System of twenty-four satellites eases the task of navigation, but the precision of the system ultimately depends on an accurate grid of star positions. Observatories in space, such as the Hubble Space Telescope, also require guide stars to point at a specific target.

Star charts, even those in computer data banks, need constant updating because the stars are moving in space. Hipparchos, the European Space Agency's astrometric satellite, recently measured the positions of more than a hundred thousand stars, with an accuracy of a few milli-arcseconds. Within a few years, however, the nearer, brighter stars will have wandered that far from their positions. Therefore the navy is preparing to monitor the changes with its interferometer. In addition, the navy team hopes to do some basic astronomy on binary star orbits, stellar masses, and stellar diameters.

The Navy Prototype Optical Interferometer (NPOI) has had a long adolescence. It began with experiments in the late 1970s by Michael Shao and D. H. Staelin at the Massachusetts Institute of Technology. For his doctoral dissertation, Shao devised a way to track the fringe motion caused by poor seeing. Shao and Staelin built a prototype two-mirror interferometer in 1980 with a baseline of 1.5 meters that proved that fringes could be stabilized to within the width of a spider's silk thread. By counting fringes in two colors simultaneously, they could stabilize fringes for several hours.

The fringe-counting technique was incorporated into a series of arrays on Mount Wilson, funded by the navy, which culminated in 1986 with the Mark III. It had baselines of 8 and 12 meters. In 1989 this instrument was able to measure the diameters of twenty-four giant and supergiant stars with a precision of 50 micro-arcseconds. At that point the navy committed funds to build the NPOI. It proceeded with dispatch. Ground was broken on Anderson Mesa in October 1992, and the first fringes were seen exactly two years later.

The navy has spared no expense to build a state-of-the-art instrument. The astrometric portion of the array consists of four fixed siderostats arranged in a Y, with baselines between 19 and 38 meters. Lasers are used to monitor the length of the baselines with a precision of one-hundred-thousandth of a centimeter. The telescopes have 35-centimeter mirrors, a size that is a compromise between desirable collecting area and undesirable blurring due to seeing. The designers hope to be able to measure fringes on stars brighter than magnitude 8.

The imaging part of the NPOI has six siderostats that can be stationed at any of thirty positions within a Y. Baselines between 2 and an impressive 437 meters are available. The 12-cm telescopes are small enough to avoid the worst of seeing effects, but tip-tilt mirrors and vacuum pipes for the light beams have also been installed.

In recent years the navy team has been demonstrating the power and versatility of the NPOI. Zeta Orionis, previously thought to be a single star, turned out to have a faint companion at a distance of 42 milli-arcseconds. The disks of fifty bright stars such as Delta Cephei and Eta Acquilae have been measured with sufficient resolution to detect a variation of brightness from center to edge. And the orbits of many binary star systems have been determined with high precision.

BRING ON THE BIG GUNS

All the optical arrays we've described so far have impressive angular resolution over a field of view of several arcseconds. Their small mirrors limit them, however, to objects brighter than about magnitude 7. Probing deeper into space will require far more collecting area than these arrays possess. The European Southern Observatory and the Keck Observatory already have the acreage of glass needed for deep-sky interferometry. So over the past decade each observatory has been building the equipment to convert its telescopes into an interferometric array. They each hope to attain a precision of less than 100 micro-arcseconds in the position of an object on the sky, which would enable them to detect the wobble of a star caused by the attraction of a planet.

The Europeans, to give them credit, had this idea from the beginning of the VLTI project in 1987. Jacques Beckers, whom we have met before, joined ESO in 1989 to spearhead their effort. He was responsible for three key decisions: first, the choice of Paranal as a site for the VLT; second, the spacing of the four 8-meter telescopes so as to form six baselines in a nonredundant array; and third, the addition of at least two meter-class auxiliary telescopes.

When the construction is complete, the four large telescopes will be able to work as a separate nonredundant array. Baselines as long as 130 meters will enable them to resolve an angle as small as 1 milli-arcsecond. Four 1.8-meter telescopes will be used, either as an independent array or in combination with a pair of 8-meter telescopes. Two 1.8-meter telescopes were installed in 2004; fringes were obtained in March 2005. These auxiliary telescopes can be moved to occupy thirty different stations, with baselines as long as 200 meters. Six optical delay lines are housed in the open air inside an underground tunnel. A facility called FINITO combines as many as three beams. It also tracks the drifts of the fringes and sends error messages back to the delay line carriages.

Forming fringes by combining the beams of 8-meter and 1.8-meter telescopes is no easy matter. It's a little like trying to harness a team of elephants and mules. It has required a decade-long struggle to devise compensating and beam-forming optics. Each level of performance has revealed new obstacles to overcome.

Mounting adaptive optics on each of the 8-meter telescopes is an essential aspect of the project. The ESO is building four copies of the Multi-Application Curvature Adaptive Optics (MACAO), and the first system was installed in May 2003. As we mentioned in chapter 9, MACAO employs a sixty-element bimorph mirror and a curvature wave-front sensor. Each large telescope should reach its diffraction limit of resolution (15 milli-arcseconds in visible light) and detect objects one hundred times fainter than before. The seeing is supposedly good enough in the infrared at Paranal to allow the 1.8-meter telescopes to operate with only a tip-tilt correction to the image.

The VLTI is a huge project that has been proceeding slowly in stages. As each new component is added and each milestone is passed, some new science emerges. The first fringes were obtained on March 17, 2001, from Sirius. On the next night the diameter of a star (Alphard) was measured at 9 milli-arcseconds, the angle between a car's headlights at a distance of 35,000 km. In September 2002 the beams from all pairs of the 8-meter telescopes were combined for the first time.

And in November 2002 the diameter of the nearest star, Proxima Centauri, was measured. It is a red dwarf star, whose mass is only one-tenth that of the Sun. Even at a distance of 4.2 light-years, it has a magnitude of 11 and would be a challenge for any array such as COAST. The VLTI easily measured its diameter, however, as one-seventh that of the Sun.

Although we tend to think of stars as nearly perfect spheres, they must be somewhat flattened because they rotate. The VLTI found an extreme example of flattening in June 2003. Achernar, the tenth brightest star in the sky, has an equatorial diameter 50 percent larger than the polar diameter. This is a real puzzle. Some forces not generally included in the theoretical models of rotating stars seem to keep Achernar from flying apart.

Perhaps the most exciting recent news from the VLTI is the glimpse of the violent core of a galaxy. Many galaxies have *active nuclei* that are extraordinarily bright at all wavelengths from x-rays to radio waves. Several lines of evidence suggest that a supermassive black hole in the nucleus generates the light by sucking gas, dust, and stars into a dusty envelope that surrounds the hole. Such envelopes have been detected indirectly by their heat radiation, but their sizes and shapes are a matter of debate. Are they disklike or doughnut-shaped? How do they change, and how do they generate the vast quantities of energy they emit?

In June 2003 a team of European astronomers used the VLTI to obtain an infrared image of a galactic nucleus (fig. 10.11). They resolved the inner 30 milli-arcseconds of the bright spiral galaxy NGC 1068 at a wavelength of 10 microns. At the distance of this galaxy, 60 million light-years, the core is only 10 light-years in size. So far the actual shape of the core is uncertain, but better images are on the way. With such images astronomers will be able to test their ideas of how the energy of the nucleus is generated. This was only the second time a galactic nucleus had been resolved. The Keck Observatory had observed a nearer galaxy a month earlier.

When do you suppose we can we expect the VLTI to find the first extrasolar planet from the wobble of its parent star? That will require equipping two

Fig. 10.11. NGC 1068 is a bright galaxy with an intense central nucleus. This visible-light image was obtained with the Gemini 8-meter telescope on Mauna Kea. The central portion has been magnified in the inset on the right. The small circle outlines the region the Very Large Telescope Interferometer has resolved, 30 milli-arcseconds in diameter.

or more 8-meter telescopes with a copy of the PRIMA dual-beam facility (note 10.6), which may not happen under the present funding schedule until 2007. In addition, the four 1.8-meter telescopes must be phased into the array of 8-meters. The first of these smaller telescopes was added in January 2004, the second in December.

Speaking of funds, the projected cost of the VLTI through 2008 is about EUR 100 million.

THE KECK INTERFEROMETER

Extrasolar planets are big news in astronomy. The public is fascinated by the possibility of life outside our solar system, and the best places to search are planets similar to Earth and in similar orbits. Since 1995 more than a hundred Jupiter-sized planets have been found orbiting nearby stars, but many are too close to their parent star to support life. NASA is betting that the Keck Observatory can be the first to find more favorable candidates. As part of its Origins program, NASA has funded the construction of the Keck Interferometer (KI). This world-class instrument will combine the two 10-meter giants on an 85-meter baseline with four auxiliary telescopes, on baselines as long as 135 meters.

The KI draws on the experience gained in a pilot program of astrometry at the Palomar Testbed Interferometer between 1995 and 1998. The project was funded by NASA and carried out jointly by the Jet Propulsion Laboratory (JPL), Caltech, and the University of California. To search for a planet orbiting a star, the team learned to monitor the position of the star relative to another nearby star that is not suspected of having a planet. This kind of *differential* astrometry is far less sensitive to seeing than a measurement of the position of a single star. A wobble in the position of a star as small as 30 micro-arcseconds might be detectable in an hour-long observation. That would allow the KI to discover a planet the size of Uranus at a distance of 60 light-years.

Construction of the KI began in 1998, and the first fringes were obtained with the two 10-meter telescopes in March 2001. Most of the components of the Keck Interferometer and the VLTI are similar. They both have an adaptive optics system, a dual-star module, optical delay lines, tip-tilt correction, and a fringe tracker.

Both interferometers also have a novel device called a *nulling beam combiner* (ESO's gadget goes by the name GENIE). By introducing an appropriate phase delay, the observers can cancel the blinding light from a central star and allow only the faint light from its circumstellar disk or planet to enter the beam combiner. This device will be a tremendous aid in searching for planets. It works best at infrared wavelengths as long as 10 to 30 microns, where a cool planet is "only" a few million times fainter than its parent star.

The design of the KI calls for four to six auxiliary telescopes, each 1.8 meters in diameter and housed in its own small dome. But as we read in chapter 4, native Hawaiians are wary about further construction on Mauna Kea, a site they consider sacred. The State Office of Hawaiian Affairs sued NASA in 2002 to show the full environmental impact of the proposed construction. The federal court ruled that a revised Environmental Impact Statement must be prepared and accepted. The issue remained unresolved as of April 2005.

As of March 2005 the KI hasn't discovered a planet, but it has made possible a lot of interesting science. In October 2002, for example, a team led by Mark Colavita of JPL measured the diameter of a circumstellar disk around the star DG Taurus. It is an example of a very young Sun-like star that may be forming a planetary system. The disk has an inner radius of one-tenth of the distance of the Earth from the Sun, rather close to the star for comfortable living.

Then in May 2003 a large team led by Mark Swain of JPL made a real coup. They used the two Keck telescopes as an interferometer to resolve the violent nucleus of a galaxy, NGC 4151, at a distance of 40 million light-years. They determined that the nucleus is no larger than 0.3 light-year, a result that rules out some models of the accretion disk around a black hole. Four weeks later a team at the VLTI achieved a similar result on a more distant galaxy, Messier 77.

The future looks bright for these impressive interferometers. We can look forward to more discoveries and perhaps some solutions to old problems. Astronomers have proven the adage that two eyes (or more?) are better than one.

TOWARD THE EVER-RECEDING HORIZON

In March 2004, NASA published a portrait of the universe as it looked thirteen billion years ago. This was not some artist's conception, but the genuine article. The Hubble Space Telescope had stared at a tiny patch of sky for a *million seconds* until it had collected enough light to form an image. This keyhole view of the universe, the Ultra Deep Field, contains about ten thousand galaxies. Many are familiar spiral and elliptical galaxies. But among the most distant galaxies lies a collection of strange objects (fig. 11.1). They look like knitting needles, necklaces, or bits of twisted rope. Some have no regular shape at all. Astronomers at the Space Telescope Science Institute who analyzed the image think they have discovered the building blocks of the first galaxies. They may have formed only a few hundred million years after the Big Bang.

Astronomers are intensely interested in how the first galaxies were assembled. What caused the cold gases to collapse? What were the masses of the building blocks? When did the first stars appear? How did these young galaxies evolve after the stars lit up? How, in short, did the structure that we see everywhere in the universe originate?

To find answers to such questions astronomers will need high-quality spectra of these faint galaxies and the gas around them. There is a problem, however. The Hubble, with its 2.4-meter mirror, required an exposure of twelve full days to capture this single image. Spectrographs spread out the available light into all its constituent colors, so high-quality spectra would require much longer exposures. The HST, despite its power, is ill suited to this task.

Big ground-based telescopes can help out. Indeed, over the past decade

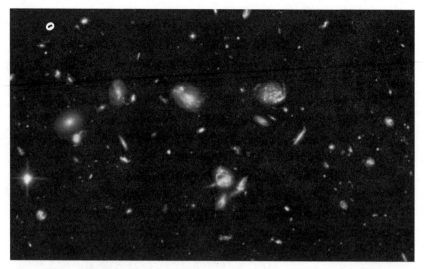

Fig. 11.1. Part of the Hubble Ultra Deep Field. The image shows galaxies from the present back to when the universe was less than a billion years old.

much of the spectroscopy needed to follow up on the HST's discoveries has been done from the ground. The 8- and 10-meter giants have been particularly effective. But the HST has continually raised the ante, and ground-based telescopes can't keep up with it.

In 1995 the Hubble's first Deep Field images gave us a glimpse of the universe at an age of about a billion years. In 2004 the HST has looked back farther in time and deeper in space. At these distances normal galaxies are exceedingly dim as well as heavily redshifted. (Quasars, by comparison, are unusually bright and much easier to detect.) So the present generation of telescopes is just not up to the task of studying these very young objects. That may be hard to accept, but it is true. By pushing deeper into space and finding this faint wonderland in the early universe, the HST has once again demonstrated the need for more light.

The evolution of the early universe is only one intriguing subject that is pushing the envelope. We would like to study the formation of stars and planetary systems in distant galaxies. We want to identify the dark matter that accounts for most of the universe's mass and the dark energy that causes space to expand ever faster. And we would like to find distant planets where life as we know it might survive.

Fig. 11.2. The Euro50 telescope would stand 85 meters tall. It is the second largest telescope being proposed at present.

Some enterprising astronomers have therefore proposed to build telescopes of staggering dimensions, with mirrors 30, 50, and even 100 meters in diameter. Figure 11.2 shows a concept for one of these so-called extremely large telescopes (ELTs), the 50-meter Euro50, which at 85 meters tall would dwarf a Boeing 747. Imagine what a 100-meter would look like. And imagine, if you can, the exciting science such Goliaths could bring within our reach.

Building a new telescope generally takes ten to fifteen years, so if you want one you'd better start thinking about it well in advance. In North America and Europe, half a dozen groups began their planning several years ago. They already have fairly detailed designs and are beginning to search for funds. Money is a big problem, however. Even the cost of developing the necessary technology is daunting. The huge price tags on ELTs is encouraging everyone to look for partners. Despite the technical obstacles and the costs, a few ELTs may see first light within a decade.

That would be none too soon. As we shall see, NASA is already building the successor to the HST, a 6.5-meter telescope with unique abilities to see deeper into space. And an international consortium broke ground in November 2003 for the Atacama Large Millimeter Array (ALMA), a radio interferometer that will reveal the intricate structure of molecular clouds and dust everywhere in space. Both of these new observatories should be complete by 2012. Although they will be unique in some ways, they, like the HST, will depend on ground-based telescopes for high-resolution spectra and images. That could be one of the tasks of a new generation of telescopes.

Designing a new telescope usually starts with a survey of the critical observations it has to make, its *science drivers*. One has to clarify the scientific goals before deciding on specifications. Then scientists have to consider the technology that is available now or will be in the near future. There are questions galore. How large a mirror is needed? Could we outfit the telescope with adaptive optics? What would the collection of focal plane instruments look like? Where could we build the telescope, and what are the requirements for the site? The process is iterative; the science drives the design, the design helps to refine the science goals. In this chapter we will follow the process.

The first question is why. Why build such a colossus? What will it have to do? To understand the answers, we first need to recall some of the outstanding issues in astronomy during the 1990s.

A TOUR OF THE HORIZON

Cosmology

Astronomers have known for decades that galaxies form in clusters, but they thought the clusters were distributed randomly in space. In 1981 Margaret Geller and John Huchra, astronomers at the Center for Astrophysics in Cambridge, Massachusetts, carried out a deep survey of galaxies over a wide swath of the sky. Using redshift as a distance indicator, they were able to determine the locations of the galaxies in three dimensions. Their data revealed a gigantic void, almost empty of galaxies, and a Great Wall of luminous galaxies 500 million light-years across (fig. 11.3). Then, in 1989, clusters of galaxies were found to outline the surfaces of gigantic "bubbles" whose interiors are almost empty of bright matter. This frothy structure must have formed early in the development of the universe, but how? The familiar forces of gravity and pressure don't seem adequate.

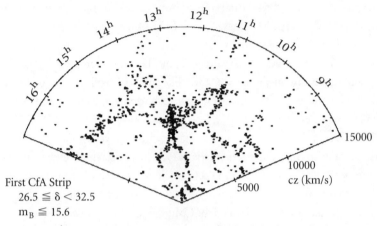

Fig. 11.3. The Great Wall is a sheet of galaxies over 500 million light-years wide that bounds a huge void. It was the first such large-scale structure discovered.

Before we can answer the question, we need much more detailed information on the distribution and masses of clusters of galaxies, within a representative volume of space. Several surveys, including the Sloan Digital Survey, the Two-Degree Field Redshift Survey, and the Keck Deep Field, among others, have been under way for a decade. A preliminary picture is building up, out to a modest redshift of $z = 1$, which corresponds to an age of six billion years. One important task for the next decade will be to obtain an even larger sample at greater distances. And that will require telescopes with much greater collecting area and wide fields of view. The 8- and 10-meter telescopes around the world will be occupied with this work for much of the coming decade and will be able to push out to perhaps $z = 4$. Beyond that, the ELTs must take over.

Dark matter has become an immensely important factor in cosmology. Its existence was deduced in the 1970s from its role in holding together a cluster of galaxies, and from the rotation profiles of individual galaxies. During the 1990s we learned that the elusive dark matter, whatever its composition, comprises over 90 percent of all the mass in the universe. At the moment the only method we have of sampling its distribution in space is based on the *gravitational lensing* effect (note 11.1).

Strong lensing of a distant galaxy by a nearer galaxy has allowed astronomers to determine the masses of some galaxies. *Weak* lensing now produces the first maps of the elusive dark matter. Several groups have analyzed the distortions that dark matter introduces in the images of clusters of galaxies. A U.K.-German group observed a pair of clusters, Abell 901/902, that lie quite near, at $z = 0.16$. The dark stuff seems to overlay the luminous matter fairly closely, and it also connects different clusters across great dark spaces (fig. 11.4). These observations tend to support predictions that the dark matter has a highly filamentary structure. This is an exciting field that is just opening up. To probe more deeply, the distortions of much fainter galaxies will have to be imaged. An ELT could excel at this task.

For many years astronomers thought the rate of expansion of the universe, expressed by the famous Hubble constant, was the same at all distances. Two observational methods gave discordant results for the constant, but in the 1990s its value converged to a single number (57 km/s/megaparsec), uncertain by perhaps 10 percent. Then in 1997 we learned that the rate of expansion of space is *accelerating* at great distances. The Hubble constant is not the same everywhere, which suggests that a repulsive force exists, a so-called dark en-

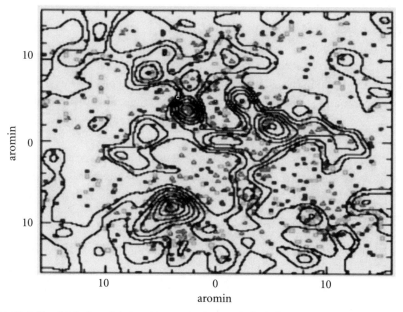

Fig. 11.4. The distribution of dark matter near luminous galaxies is shown as contours in this map of the cluster Abell 901/902. Galaxies are represented by small symbols: triangles are spirals, squares are elliptical galaxies. Notice how the galaxies congregate near regions of high dark-matter density. A tongue of dark matter *(lower center)* connects the two clusters.

ergy whose nature is totally unknown. The discovery of dark matter and dark energy has left astronomers mystified and intrigued. They will need new observations to identify their physical nature.

Galaxy Formation

Astronomers have made great progress in understanding the evolution of galaxies, but not as much in understanding their formation. We learned that galaxies grow by devouring their neighbors. (Our Milky Way is dining on a dwarf galaxy even now.) Galaxies were observed to collide, merge, and tear their neighbors apart through tidal forces. We saw how successive generations of stars could change the chemical distribution and internal motions of galaxies. We learned that the rates of star formation vary tremendously in galaxies of different types and ages, but we didn't understand why.

In 1995 the Hubble Space Telescope obtained its first Deep Field images. They showed well-formed galaxies at distances up to 13 billion light-years. These young galaxies must have been born in the first billion years after the Big Bang. How was this possible? Astronomers were left with a long list of questions regarding the formation of galaxies.

What triggers the collapse of a cold, dark gas cloud? How do the bits and pieces that form in the cloud draw together to create a proto-galaxy? How are the masses and spins of galaxies determined? Then, after a proto-galaxy is born, how does it evolve? How do spiral and elliptical galaxies develop from early forms? At what age did the first stars begin to shine? How does the distribution of stellar types change within a galaxy? How does the chemical composition of a galaxy change as old stars die? Answering such questions will require spectroscopy and imaging of faint objects, which would take unreasonable amounts of time with the present generation of telescopes. Here is a central task for the ELTs.

Then there are the intriguing black holes. At some stage in the development of a galaxy, a massive black hole begins to form in its nucleus. We have guessed that the most distant quasars, at $z = 5$ and 6, contain supermassive black holes that generate the quasars' extraordinary emissions. More recently black holes have been found and weighed in nearer galaxies. Black holes with smaller masses are also abundant. But we are left with the question of how such holes originate. Here we need the milli-arcsecond angular resolution that a diffraction-limited ELT could provide.

Star Formation

One of the great triumphs of twentieth-century astronomy was the explanation of how stars evolve during their lifetimes. Observations of star clusters like the Pleiades had shown that stars heavier than the Sun evolve rapidly, changing color and brightness in a precise pattern as they age. Less massive stars, like our Sun, remain relatively unchanged for five billion years or more. The theory clarified why these changes occur and then led to calculations that matched the observations nicely. The theory also explained how stars manufacture heavy elements. When a star explodes as a nova, it sprays its heavy elements into the surrounding gas. From these cold gases, new stars are born with enriched composition.

What was lacking in this beautiful theory was a physical description of how a star forms from the interstellar gases. What processes cause a gas cloud to cool and contract? What is the role of the interstellar dust? Why are stars born in groups more often than singly? What determines the distribution of masses in a cluster?

Back in 1983 the Infrared Astronomical Satellite (IRAS) had mapped the sky in infrared light and shown that some distant galaxies were experiencing tremendous bursts of star formation. With the launch of the Hubble Space Telescope in 1990 and the publication of its amazing images (e.g., fig. 11.5), star formation took center stage in astrophysics. New infrared detectors enabled astronomers at ground observatories to pierce the thick curtains of dust within a giant molecular gas cloud and observe the signs of star birth in the Milky Way. Their infrared spectrographs revealed high-speed jets and winds. Accretion disks were discovered that could evolve into full planetary systems. Star formation turned out to be highly variable and difficult to unravel. Many more examples would be needed, especially in other galaxies, and finer details will have to be observed before some clarity could prevail. Here was another task for the ELTs.

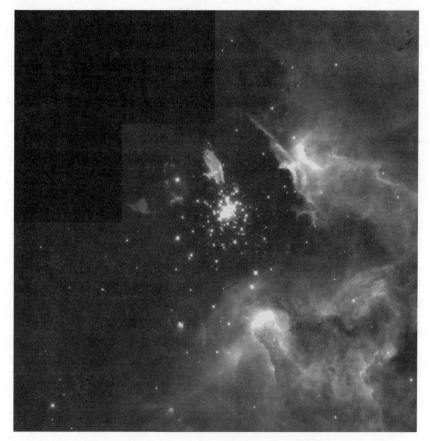

Fig. 11.5. NGC 3603 is a nebula in the Milky Way that contains several star nurseries. They appear in the right half of this image, which was obtained with the Hubble Space Telescope.

Planets

We have long wondered whether we are alone in the universe. Does life exist outside our own solar system? Since the 1960s Frank Drake and his band of dedicated radio astronomers have searched without success for intelligent life in the universe (note 11.2). We would be delighted to find signs of any type of life, even microbes. At the time of this writing, two robots were scouring the sands of Mars to find signs of water, the essential ingredient for life.

Another way to search for extraterrestrial life is to find habitable planets. For many years attempts to discover an extrasolar planet failed. Then, in 1995, two Swiss astronomers used a clever technique on a small telescope to discover the first planet orbiting a nearby star, 51 Pegasi. Astronomers scrambled to the Keck, Lick, and Palomar telescopes to find more. By the end of the millennium over a hundred had been found, a few in the same planetary system.

Most of these planets were giants, similar in size to Jupiter, and orbited very close to their host stars. None was likely to support life. Can we find planets similar to Earth, in favorable orbits? We will have to examine more distant stars and be able to detect much smaller planets. The search will require larger telescopes, with more collecting area and extraordinary angular resolution. If we find some promising candidates, we would hope to examine their atmospheres for signs of oxygen, water, or methane.

THINKING BIG

As the twentieth century drew to a close, astronomers were realizing that settling many of the outstanding issues would require more light and sharper images. Advancing the science was their prime motive for building a new generation of telescopes. A second obvious motive was the prospect of discovering something entirely new, a breakthrough of some kind. In the past each increase in the power of a telescope led to major discoveries. (Recall Edwin Hubble and his discovery of the recession of the galaxies, made possible with the completion of the 100-inch, or 2.5-meter, telescope.) And finally there was the success of the Keck telescopes. They had opened a new path toward telescopes larger than 10 meters.

Jerry Nelson probably started to think about a 25-meter successor to the Keck telescopes soon after they were completed, around 1997. He was not alone. The success of the segmented mirror design inspired more than half a dozen groups to begin to plan bigger telescopes:

- At the Gemini Observatory, Jeremy Mould and Matt Mountain were proposing a Maximum Aperture Telescope (MAXAT), in the range of 25 to 50 meters.

- A group of Swedish astronomers at Lund Observatory led by Arne Ardeberg proposed a 20-meter in the 1980s, then raised their sights to 30 meters and finally aimed at a 50-meter, the Euro50. They had their eye on the Canary Islands as a potential site.

- Canadian astronomers at the Association of Canadian Universities for Research in Astronomy (ACURA) were aiming at a more modest 20-meter, the Very Large Optical Telescope (VLOT), that could replace the Canada-France-Hawaii Telescope on Mauna Kea.

- A group of American universities was proposing the Large Atacama Telescope, a 25-meter telescope optimized for submillimeter radiation. Riccardo Giovanelli, a professor at Cornell University and a radio astronomer of some note, was the project leader.

- The consortium that built the 6.5-meter Magellan telescopes at Las Campanas was now proposing the Giant Magellan Telescope (GMT), with a segmented 25-meter mirror. Roger Angel at the University of Arizona was their lead designer.

- The builders of the Hobby-Eberly Telescope were also thinking about a 25-meter successor to HET.

- The most ambitious group of all, led by ESO astronomer Roberto Gilmozzi, was planning the 100-meter Overwhelmingly Large Telescope (OWL).

These groups all had similar scientific objectives but varied considerably in their proposed hardware. Between 1997 and 1999 many of them presented their ideas at half a dozen international conferences. The Swedes, for example, hosted a conference on ELTs at Backaskog Castle in 1999. Two meetings were held among the potential users of Gemini's MAXAT. NASA sponsored several meetings to discuss the complementary roles of the ELTs and the James Webb Space Telescope. Such conferences helped to firm up ideas on the potential science and the technical challenges of an ELT. They also focused attention on what an ELT could *not* do well.

FINDING YOUR NICHE

More and more astronomical observations are made at infrared wavelengths. The cold gases in space, the icy objects at the edge of the solar system, and the warm cocoons around young stars are best observed at wavelengths beyond 1 or 2 microns. Extragalactic astronomers also prefer to observe in the infrared because the galaxies of interest are strongly redshifted. As an extreme example, the young galaxies discovered in the Ultra Deep Field lie at shifts of $z = 6$

to 12. Their visible light has shifted into the *thermal infrared,* at wavelengths longer than about 10 microns. And that raises a problem. The Earth's atmosphere is nearly opaque at such long wavelengths.

In figure 11.6 we see the transmission of the atmosphere as it varies with wavelength. Water vapor, carbon dioxide, methane, and ozone molecules block the light in a series of bands between 1 and 30 microns. Some useful windows appear between 1 and 5 microns and around 8 to 12.5 microns. Beyond 20 microns only a few narrow windows are open. To make matters worse, the warm atmosphere radiates its heat in the thermal infrared. As a result, the sky looks very bright at long wavelengths. A ground-based telescope has to look through a dazzling foreground, like a bright fog, to detect a distant faint galaxy.

Here lies the great advantage of space telescopes. They can orbit far above the atmosphere, untroubled by its opacity, its emission, or its turbulence. NASA was fully aware of these virtues when planning the successor to the HST, the James Webb Space Telescope (JWST). It was designed specifically to observe the early universe at long wavelengths (note 11.3).

As it stands now, the JWST will have a lightweight beryllium mirror, consisting of eighteen segments that will open up in space to a 6.5-meter diameter (fig. 11.7). That is small even by current standards on the ground, but it still represents a sevenfold increase in collecting area over the HST. The JWST will be optimized for observations at 2 microns, where it would have a diffraction limit of 0.06 arcsecond. It will chill down to about 50 kelvin, a temperature at which its own thermal emission should be negligible. A trio of cameras will cover the wavelength range from about 1 to 30 microns.

The JWST will be the ideal instrument to find objects at wavelengths longer

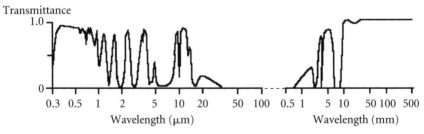

Fig. 11.6. The atmosphere is transparent to infrared radiation only in a series of windows between 1 and 20 microns, and again beyond 1 mm.

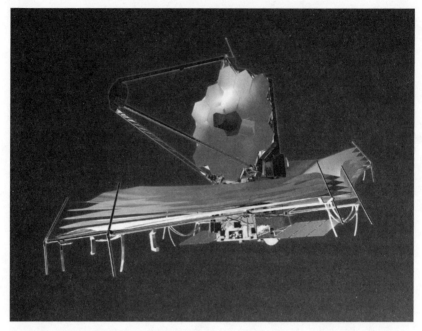

Fig. 11.7. An artist's conception of the James Webb Space Telescope. In this rendition the 6.5-meter mirror has eighteen segments.

than, say, 10 microns. Its spectrograph will have only modest wavelength resolution, however, with limited ability to analyze the faint objects it finds. That is the JWST's weak spot, which opens an opportunity for an ELT to help out.

NASA plans to launch this $2 billion gem into an orbit at the L2 Lagrangian point (note 11.4) around 2012. Like the Hubble, the JWST would be available to any qualified astronomer with a good observing program.

Now, if you refer to figure 11.6, you will see that the atmosphere's transparency improves at wavelengths longer than 1 mm. This is the spectral region where the Atacama Large Millimeter Array will shine. Many of the simple molecules in space, such as carbon monoxide and water, absorb and emit light at millimeter wavelengths. So ALMA will enable us to study the composition and motions of the cold, dense clouds that collapse into galaxies, stars, and planets. It could map the frothy large-scale structure of the intergalactic gases and would probably do a better job of this than an ELT could at optical and infrared wavelengths.

Water vapor in the Earth's atmosphere is the biggest obstacle to observing at millimeter wavelengths. Therefore ALMA will be located at one of the driest places on the planet, Chile's Atacama Desert, at an altitude of 5,000 meters. It will consist of sixty-four 12-meter antennas spread out over a desolate plateau. With baselines as long as 10 km, this interferometer should resolve objects as small as 10 milli-arcseconds. And with a collecting area of 10,000 square meters, it will be able to detect point sources twenty times fainter than the Very Large Array in New Mexico.

ALMA is a large international project, with funding partners from Canada, Chile, Europe, Japan, and the United States. Collaborating American institutions include Caltech, Harvard, and Cornell. Scores of top-flight scientists will be able to access this magnificent instrument, although working in the thin air at 5,000 meters will be challenging.

Ground was broken for the new array in November 2003. If the present schedule holds up, ALMA will go on the air in 2012, about the same time as the JWST.

There is no doubt that the JWST and ALMA would be the instruments of choice for observations at wavelengths longer than, say, 10 microns. A ground-based ELT, on the other hand, would have an overwhelming advantage at wavelengths shorter than about 5 microns, because of its huge collecting area and, with adaptive optics, its milli-arcsecond angular resolution. For many types of observations the three types of telescopes could support each other.

Take the most severe case, the universe at an age of four hundred thousand years and at a redshift of $z = 12$. The JWST could readily *find* faint objects at a wavelength of 10 microns, but it is not equipped to obtain high-resolution spectra. Given the objects' coordinates, an ELT could observe their ultraviolet spectra, which would be shifted from 0.2 micron to an accessible 2.4 microns. ALMA could provide observations of the local gas composition, density, and temperature. And for nearer objects, such as galaxies in our local cluster, the JWST and ALMA could work together even more effectively.

HOW TO BUILD A BEHEMOTH

Once the scientific goals for a new generation of telescopes were clear, the designers could get started. At a meeting in 1999 Jerry Nelson and his colleague Terry Mast surveyed many of the issues involved in the design of an ELT. With their experience of building the Keck telescopes, they naturally focused on telescopes with large segmented mirrors. Although in principle a segmented mirror could be built to almost any size, they cautioned against taking too large a step beyond 10 meters. Experience suggested that a step of a factor of two to three would be conservative, in terms of cost, complexity, and schedule. In contrast, an immediate jump to 100 meters, as ESO was proposing, could involve severe technical risks.

Three other issues stood out in Nelson and Mast's survey: adaptive optics, overall cost, and schedule. First, they had to admit that adaptive optics for a 30- or 50-meter telescope was beyond the existing technology. With many correcting elements in the system, real-time computing speed would be only one of several hurdles. Exciting science could be done, however, using the natural image quality at a good site. Many groups were working on adaptive optics, and there was good reason to expect that within the ten- or fifteen-year lifetime of the project, the technology would improve. In any case, adaptive optics for an ELT would not be cheap.

As for the overall cost, experience suggested that the cost of a telescope quintuples if the mirror size is doubled. Thus if the Keck design were naively scaled up to 25 meters, the cost could rise to $1 billion. Such a sum would be difficult if not impossible to find, at least for a group of academic astronomers. This reality forces telescope designers to find some effective cost-cutting measures.

Nelson and Mast also emphasized the narrow window of opportunity. NASA planned to launch the JWST around 2012. ALMA would begin observing about the same time. Therefore, the design of an ELT had to start immediately, to ensure that it would be ready in time to work in tandem with these world-class facilities.

Finally, Nelson and Mast presented a "straw man" design for a 25-meter to clarify the issues involved in building an ELT. Even at this early stage, the cost of fabricating the segmented mirror was a big worry. The size and number of segments were therefore a critical choice. The Keck segments were hexagons measuring 90 cm on a side. A larger segment would bend more under gravity, would deviate more from a sphere and therefore be harder to polish, and would be more sensitive to positioning errors. Smaller segments avoid these pitfalls but require more actuators, fancier software, and more complex alignment hardware. In the end, Nelson and Mast chose a hexagonal segment 50 cm on a side. The complete mirror would then consist of 756 segments, with the central 19 missing to permit the light to pass to the instruments.

To cut fabrication time and costs, Nelson and Mast proposed to polish several small segments simultaneously. Each one would be stressed with a mechanical fixture and polished to a spherical surface. When the stress was relieved, the segment would spring into the desired nonspherical shape. All the problems involved in figuring and testing the Keck segments would return with tighter tolerances, but they would be faced with the confidence that comes with experience.

FOLLOW THE MONEY

As the new millennium approached, each group interested in building an ELT continued to develop its conceptual designs. Everyone soon realized how complicated the task had become. One can't design an ELT without taking into account a bewildering variety of interlocking factors. The type of telescope optics, the size and complexity of the next generation of science instruments, the quality of the site, the optimum wavelength range, and the constraints of the adaptive optics system are all important.

The immediate problem, however, was money. Where could they find the money to get started? Where would they ever find the hundreds of millions needed to actually build their dream machines? The European Southern Observatory, owner of the VLT, is a huge multinational institution and has deep pockets. If the OWL proves to be technically feasible, ESO could presumably find the necessary funds. The Euro50 group, with Sweden as its leader, could also count on multinational support, although with less certainty.

In the United States the competing groups divided into two camps, one favoring private sources and the other public funding. Caltech and the University of California decided early on that they would seek private funds exclusively for their 30-meter, the California Extremely Large Telescope, or CELT. They had been successful with this approach in building the Keck telescopes and had retained nearly all the observing time as a result. A board of directors, a scientific steering committee, and several working groups were set up at Lick Observatory. Jerry Nelson assumed the role of project scientist.

The Large Atacama Telescope consortium, led by Cornell University, also favored private funding. The Giant Magellan Telescope group had the wealthy

Carnegie Corporation in its corner but would have to find additional private funds. Neither of these groups was totally averse to seeking federal funds but preferred private money, because it would give them total control of the final facility.

On the other hand, the National Optical Astronomy Observatory and the Gemini Observatory were heavily dependent on U.S. federal funding. Their concept for MAXAT might attract foreign partners, but a large share of the necessary funds would have to come from the government. And by a fortunate coincidence the government was, at that very moment, awaiting a crucial report on the funding of astronomy.

THE EVENT OF THE DECADE

The National Science Foundation (NSF) and the National Aeronautics and Space Administration (NASA) are the major funding agencies for astronomy in the United States. Once a decade they seek independent advice on which projects to fund, in what order and at what pace. For the past forty years these agencies have delegated the task to the National Research Council, an arm of the National Academy of Sciences, which organizes an Astronomy and Astrophysics Survey.

A dozen teams of experts, drawn from the nation's science and technology centers, try to assess the future directions of astronomy. Where is the science headed, and what are the opportunities for great research? What new facilities are needed to pursue these goals? The survey's purview includes both ground-based and orbiting observatories.

For a year or more the survey members hear presentations and debate the alternatives. They draw on the advice of some of the best minds in the field. Finally, after closed-door debate, the survey members issue their list of recommended priorities. The funding agencies are not bound to fulfill the recommendations. They have to fold the requirements into their projected budgets, which are uncertain from year to year. But the list does provide a long-term guide, a sort of road map.

The most recent survey was co-chaired by Joseph Taylor (a 1993 Nobelist in physics) and Christopher McKee (astrophysicist and member of the National Academy of Sciences). The fifteen members of the panel included astrophysicists, observational astronomers, theorists, telescope designers, and an administrator or two. Their report, entitled *Astronomy and Astrophysics in the New Millennium,* was published in 2001.

As its first priority in space astronomy, the members endorsed the Next Generation Space Telescope, later renamed the James Webb Telescope. The top priority on the ground was a 30-meter Giant Segmented Mirror Telescope (GSMT). The report noted that MAXAT and CELT were similar in concept and could evolve as a joint project to build the GSMT with a combination of private and public money.

No real surprises here after all. Astronomers with plenty of clout and credibility had been lobbying for their favorite projects for several years. Nevertheless, the survey's recommendations carry a lot of weight, not merely for NASA and the NSF but also for potential donors of private funds. If a group of independent experts says a project is worth doing and may be feasible, a private foundation is more likely to part with the money.

FIRST STEPS IN THE UNITED STATES

With the endorsement of the GSMT concept by the Astronomy and Astrophysics Survey, NOAO and the Gemini Observatory were ready to proceed. In January 2001 their parent organization, the Association of Universities for Research in Astronomy (AURA), organized a joint New Initiatives Office (NIO) to plan a 30-meter telescope. This publicly funded facility would be available to all qualified astronomers, unlike the privately funded telescopes. By the same token, all qualified astronomers might expect to have a voice in the design of the telescope. NIO invited comments from American astronomers but also moved ahead rapidly with its in-house engineering staff. That proved to cause some friction, as we shall see.

Eight working groups were established, with members drawn from universities, industry, and foreign institutions. Among other tasks, they had to draft the scientific objectives, define the telescope performance characteristics that would satisfy those objectives, and identify critical technical problems.

Eighteen months later the NIO published a so-called point design, the result of a series of studies exploring technical issues, risks, and costs. Figure 11.8 shows one concept for the GSMT at this stage. Many features of the compact telescope mount, including the position of the main axes, were copied from successful radio telescopes. The paraboloidal primary mirror has a fast focal

Fig. 11.8. An early conceptual design of the Giant Segmented Mirror Telescope, a project of the New Initiatives Office of the Associated Universities for Research in Astronomy

ratio of f/1.0 and is composed of 618 hexagonal segments, each 1.2 meters across flats. The "adaptive" secondary mirror would not only direct the light beam to the different Nasmyth foci but would also act as a part of an adaptive optics system.

The NIO also sketched out the requirements for a battery of science instruments, the criteria for a suitable site, a design for a telescope enclosure, and many other details. The main purpose of the point design, however, was to identify the technologies that had to be developed before the GSMT could be built and to explore partnerships that could undertake the work. ESO, engaged in designing the 100-meter OWL telescope, was one possible partner. Adaptive optics headed the list of critical items. The effect of wind on the telescope was a close second.

NIO planned to refine the point design to create preliminary and then detailed designs, over the period from 2003 to 2006. If that schedule held, construction might begin as early as 2008 and science operations by 2012, just in time to work with the James Webb Space Telescope and ALMA.

KECK'S BIG BROTHER

Meanwhile, Caltech and the University of California were busy designing the California Extremely Large Telescope. In February 2000 the goal was upgraded from a 25-meter to a 30-meter, in line with the recommendations of the Astronomy and Astrophysics Survey's report. Jerry Nelson was the telescope's godfather, and he had plenty of help. Working groups were set up for the telescope structure, the mirror, the adaptive optics system, the enclosure, the site, the instruments, and, not least, the science. The project was based at the Lick Observatory's offices on the Santa Cruz campus.

The schedule called for three stages: a conceptual design phase, lasting twelve months; and a preliminary design phase and a detailed phase, each lasting eighteen months. By September 2004, in short, complete plans for building the telescope could be in hand, if money could be found quickly enough. The CELT team was racing to be first.

Between February 2000 and June 2002, the working groups pumped out a flood of technical studies on every aspect of Nelson's conceptual design. The telescope (fig. 11.9) would have Ritchey-Chrétien optics with a hyperbolic primary mirror nine times the area of the Keck telescope, an adaptive secondary mirror, and a third mirror to bring the beam to instruments at the Nasmyth foci. The focal ratio of the primary is a conservative f/1.5.

At a good site and even without adaptive optics, the telescope could form stellar images as small as half an arcsecond. A 20-arcminute field of view, valuable for mapping the distribution of galaxies, would be available at a final focal ratio of f/15. For technical reasons, the telescope has neither a prime focus nor a Cassegrain focus, but only two Nasmyth foci.

As Terry Mast had guessed back in 1999, mirror segments smaller than those used in the Keck telescopes would be cheaper to fabricate. For this conceptual design, the team settled on a hexagonal segment that fits inside a 1-meter cir-

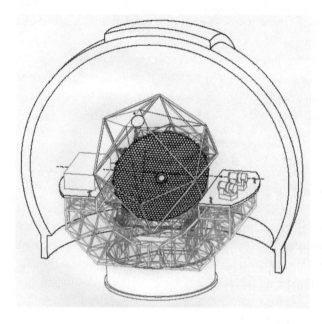

Fig. 11.9. The
proposed California
Extremely Large
Telescope

cle. The mirror has 1,080 of these smaller segments, all aligned and supported
with active controls. With a thousand segments, the CELT mirror would require sixty-two hundred sensors and thirty-two hundred actuators. Each segment must be aligned with its neighbors to within 20 nm, or about one-ten-
thousandth the width of a period on this page. The control system required to
keep this orchestra in tune boggles the mind.

By June 2002 the CELT team had generated a two-volume Green Book that
summarized the tasks for the second phase of the project. Every aspect of the
design would be reexamined to reduce risk and lower cost. Prototypes of the
control system and the stress-polishing fixtures would be built. Sample mirror segments would be fabricated from start to finish, and testing methods
would be tried out.

Wind loading on the telescope structure and the primary mirror was a real
concern. A dome 84 meters in diameter had been included in the design, but
the effects of winds on the whole assembly had to be modeled.

The spectrographs and imagers that would work with the telescope needed
to be planned in tandem with the telescope. With a field of view of 20 arc-

minutes at the Nasmyth foci, an image would be 2.6 meters in diameter, which would require instruments of unprecedented size.

Adaptive optics for the CELT was the biggest worry. To meet some of the science goals, the telescope had to deliver diffraction-limited performance— say, 7 milli-arcseconds at 1 micron. In 2002, however, no AO design was in hand, none of the major components of a conceptual system could be built or purchased, and no software for modeling a system this size was available. A major development program would be needed. Nelson, as the director of the Center for Adaptive Optics, was well aware of the magnitude of the challenge.

THE GIANT MAGELLAN TELESCOPE

Roger Angel had his own ideas on how to build a Giant Segmented Mirror Telescope, as called for by the decadal Astronomy and Astrophysics Survey. His Mirror Lab at the Steward Observatory had spun out the 6.5-meter mirrors for the Multi-Mirror Telescope and the two Magellan telescopes in Chile. He had cast the 8.4-meter mirrors for the Large Binocular Telescope as well. His mirror technology was tested and affordable. If he were called upon to design a segmented mirror equivalent in area to a 22-meter, he would start with such giant mirrors, not the meter-class bits that others were considering.

Angel's home institution, the University of Arizona, is a member of the powerful consortium that built the Las Campanas Observatory in Chile. The other partners include the Carnegie Observatories (Mount Wilson, Palomar, and Las Campanas), the Center for Astrophysics (Harvard University and the Smithsonian Institution), MIT, and the University of Michigan. This combination of private and public institutions commands impressive resources, both in funds and in brainpower.

After the second Magellan telescope saw first light (2002), its sponsors began to plan for the Giant Magellan Telescope (GMT). In the high-stakes game of deep-sky astronomy, one has to keep moving or one is soon left behind. By mid-2003 Angel and his team had produced the compact design shown in figure 11.10.

This is a Gregorian telescope (see fig. 1.3) that consists of six 8.4-meter off-axis hyperbolic mirrors that surround a central 8.4-meter mirror. The separate mirrors would be phased to act as parts of a single large mirror 25 meters

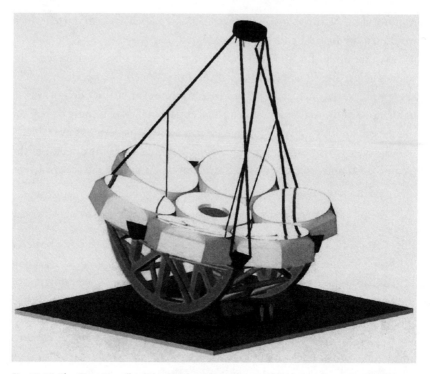

Fig. 11.10. The Giant Magellan Telescope has a mirror composed of seven 8.4-meter segments.

in diameter, as pioneered by Jacques Beckers at the original Multi-Mirror Telescope.

The beams from the seven large mirrors converge on a deformable secondary mirror. This device was developed at Angel's Center for Astronomical Adaptive Optics and was used successfully at the Multi-Mirror Telescope as part of an adaptive optics system that achieved diffraction-limited performance at infrared wavelengths. It would be part of a large system that is still in the development stage at Angel's Center for Astronomical Adaptive Optics. Even without adaptive optics the GMT could resolve 0.4 arcsecond at visible wavelengths, over a wide 15-arcminute field of view

Angel cites two important advantages of a cluster of 8.4-meter mirrors, compared with an array of meter-sized segments. Segmentation invariably introduces optical aberrations in the final image, but a few large segments would produce a cleaner image than many smaller segments. Moreover, mounting, stabilizing, and aligning large segments would be intrinsically easier than with small segments. On the other hand, polishing 8.4-meter mirrors to perform off-axis, as in the GMT design, has never been attempted.

A STORM IN ACADEMIA

By the spring of 2002, point designs had been completed for both the CELT and AURA's GSMT. Their sponsors recognized that the two concepts were quite similar and that it made sense to coordinate, if not merge, further efforts to develop the critical technology and to improve the designs. AURA and the CELT Development Corporation signed a letter of intent with this purpose in mind in June 2003. Each party would seek half of $70 million for a joint effort over the next two or three years. The CELT was renamed the Thirty Meter Telescope (TMT).

In August 2003 AURA submitted an unsolicited proposal to the NSF for $35 million, its share of the development money. And in October 2003 Caltech received the first half of a grant of $35 million from the Gordon and Betty Moore Foundation. The prospects looked good for advancing the design and technical development of the Thirty-Meter Telescope.

At that point the consortium for the Giant Magellan Telescope raised a red flag. Eighteen members of AURA, including the GMT group, complained to the National Science Foundation that the agreement between the CELT Development Corporation and AURA might violate the principle of open competition. They argued that several telescope concepts should be pursued in case any one of them proved not to be feasible. Of course, the GMT group might have submitted their own unsolicited proposal, but perhaps they felt they were not ready.

In the face of this embarrassing development, the management of AURA made a Solomonic decision: they offered to split the baby. They withdrew their original proposal to the NSF and submitted a revised one. Half of the $35 million requested would now be open for competition, with the expectation that the GMT group would submit a proposal for this money.

At the present time, the Californians are proceeding with Phase 2 of the Thirty-Meter project, while AURA and the GMT consortium await a favorable decision from the NSF. If and when the AURA proposal is accepted, AURA would join the CELT Development Corporation as a full partner in developing the telescope. Presumably the GMT folks are also hunting for private or state funds.

The present time line for the TMT calls for a conceptual design review in April 2006; a construction proposal in September 2006; a site selection review in July 2007; and earliest start of construction by January 2008. If all goes well, first light would be possible by April 2014.

The Giant Magellan project has also made rapid progress. A conceptual design review is scheduled for the fall of 2005.

NORTH OF THE BORDER

Canadian astronomers don't propose to be left behind in the scramble toward huge telescopes. They now have access to the 3.6-meter Canada-France-Hawaii Telescope and the 8.2-meter Gemini North, both on Mauna Kea. To remain competitive, they hope to be able to replace the CFHT with the Very Large Optical Telescope (VLOT), a 20-meter segmented telescope. To that end they have organized the Association of Canadian Universities for Research in Astronomy (ACURA). Ray Carlberg, from the University of Toronto, is the project scientist.

A team based at the Herzberg Institute of Astrophysics, in Ottawa, completed a point design for the telescope in November 2003. It is not a complete conceptual design, because many technical tradeoffs still have to be made. The size and shape of the segments, the polishing techniques, and the type of mechanical supports all need further work. More important, the choice of an optical design is still under discussion.

The current Ritchey-Chrétien design has an f/1 primary mirror composed of 150 hexagonal segments, each 1.8 meters from point to point. R-C optics lead to a smaller enclosure, always an advantage, but the 2.5-meter convex secondary mirror would be difficult to figure and test. In contrast, the concave secondary of a Gregorian design is easier to fabricate and might ultimately become part of the adaptive optics system. More thought is needed.

Wind loading of the primary mirror is a serious concern. Therefore the team contracted for an intensive study of alternative enclosure designs. Instead of the usual dome or cylinder, the best choice seems to be a *calotte*. This is a hemispherical structure that is sliced at an angle (fig. 11.11). The top half rotates on an inclined ring at the slice, while the whole structure rotates in az-

Fig. 11.11. The enclosure for the Very Large Optical Telescope has a novel design, the calotte. The dome is sliced at an angle, and the top half, which contains a hole for the telescope to look through, rotates on the slice. The whole dome also rotates in azimuth about a vertical axis.

imuth on a base ring. A combination of rotations allows the telescope to look at any point in the sky through a hole in the top half. In bad weather, a door covers this hole. Wind-tunnel tests of a model of the calotte are under way.

The team has planned two adaptive optics systems. The simpler system, with fewer correcting elements, would improve the raw image resolution by a factor of two or three, over a field of view of several arcminutes. It is much closer to being feasible and would immediately enable studies of large-scale clusters of galaxies. A second system that would allow the telescope to perform near its diffraction limit is further off and would require major advances in the technology.

The VLOT project report was a big step forward in Canadian efforts to advance to the big leagues of ELTs. However, the difficulty and cost of a completely independent program soon became obvious. In June 2003 ACURA signed a letter of intent to join with AURA and the CELT Development Corporation to develop the technology needed for a 30-meter telescope somewhere, possibly not on Mauna Kea. The University of California and Caltech would seek $35 million in private funds for the effort. The U.S. NOAO and the Canadian ACURA would each seek funds from their governments for an additional $35 million. The Thirty-Meter Telescope was well under way, with a target date of late 2007 for a preliminary design and with cost estimates accurate to 10 percent.

ON THE OTHER SIDE OF THE WATER

Back in 2000 the Swedish Lund Observatory had organized a consortium to build a 50-meter segmented mirror telescope, the Euro50 (see fig. 11.2). The group now includes representatives from Finland, Ireland, Spain, Sweden, and the United Kingdom. Arne Ardeberg is the program's director and is also responsible for the science agenda. Torben Andersen, a well-known optical engineer, will oversee the telescope design, and Jacques Beckers (that peripatetic innovator) will take charge of the adaptive optics program.

The project's team members have made great progress, and in 2003 they published a comprehensive report on many aspects of the design. First of all, they have chosen a proven site at La Palma, in the Canary Islands. Second, the optical design has been firmed up.

The telescope would have a Gregorian configuration with a very fast (f/0.85) primary mirror and a 4-meter adaptive secondary mirror. The primary would consist of 618 hexagonal segments, each 2 meters across. These aspheric segments are about twice the size of the CELT's and the largest that can easily be supported. The British and Finnish partners in the project have some experience with polishing such large segments, but a prototype needs to be tested before a final decision can be made. Unlike the CELT and GSMT mirrors, the Euro50 primary mirror has a hexagonal shape. As a result, an image of a star would have six symmetric arms extending from a central core, a peculiar type of aberration.

A wheel-on-track design has been adopted for the mechanical structure of the telescope (see fig. 11.2). It resembles that of the 100-meter Effelsberg Radio Telescope, which has performed well in the past. The altitude axis is hollow to allow light to reach instruments at the two Nasmyth foci.

A 50-meter mirror would make a wonderful sail, so the team has designed an active control system to keep the segments aligned at moderate wind speeds. Wind tunnel tests suggest that wind disturbances could be as important as seeing for image quality.

The Swedes and their friends seem optimistic that a suitable adaptive optics system will be feasible, at least at a wavelength around 2 microns. They plan to install one in stages, building up to a full-blown system with laser beacons. As a start, they have designed a secondary mirror to act as a deformable AO mirror with four thousand actuators. (Someone will have to build this beauty!) The wave-front sensor and reconstructor would be placed inside the cell of the primary mirror. Without adaptive optics, the telescope resolution would be only 0.9 arcsecond because of unavoidable optical aberrations; with the AO turned on the telescope could resolve 2 milli-arcseconds in the visible and perhaps 10 milli-arcseconds in the near infrared.

How do you protect a telescope this size from the weather? The team has proposed an enclosure 120 meters high, complete with air-conditioning and wind screens. This version, almost as large as Saint Peter's Basilica, would cost an estimated EUR 125 million, however, and will need some refinements.

Present estimates for the total cost of the telescope, including the enclosure and adaptive optics, exceed EUR 570 million. This price compares favorably with the projected cost of the Thirty-Meter Telescope, $600 million to $700 million. If funds arrive in time, the Euro50 could be online by 2012, when the JWST and ALMA turn on.

GOING FOR BROKE

Could one actually build a 100-meter telescope? Roberto Gilmozzi and his friends at the European Southern Observatory think so. Their concept is breathtaking. If it could be built, their Overwhelmingly Large Telescope would be the largest conceivable for many years. So far, the ESO team has generated a conceptual design for the OWL (fig. 11.12) that looks promising.

The optical design (fig. 11.13) calls for a 100-meter primary mirror with a focal ratio of f/1.42. The secondary mirror is also huge, a full 25.6 meters in diameter, and it has 216 segments. It would be supported some 95 meters above the primary mirror and would direct the beam through a central hole in the primary.

To economize on fabrication, the primary mirror would have a spherical surface, as in the Hobby-Eberly Telescope. Each of the 3,048 hexagonal segments would be identical. Spherical surfaces are easy and cheap to generate in batches with so-called planetary polishers, but even with three or four machines running day and night the job would take six years. Zerodur, ULE, or ultra-lightweight silicon carbide are possible materials for the mirror.

Like all spherical mirrors, this one would have aberrations that must be compensated for. In the present design, four mirrors would correct the image (see fig. 11.13). Two deformable 8-meter mirrors aid in the active control of the

Fig. 11.12. The Overwhelmingly Large Telescope would stand 95 meters tall. In this artist's rendering, a truck in the foreground gives some idea of the scale of this behemoth. The giant enclosure, in the rear, would slide forward to cover the telescope.

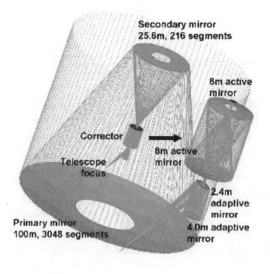

Fig. 11.13. To correct the spherical aberration of the spherical 100-meter primary mirror, the OWL has a complex optical secondary system. The flat secondary is 25.6 meters in diameter. The corrector module consists of an aspheric 8-meter tertiary mirror, an aspheric active quaternary mirror, a 4-meter adaptive mirror, and a 2.4-meter adaptive mirror. At the focus, the telescope has a 10-arcminute field of view.

segments; a 4-meter mirror focuses the telescope and a flat 2.4-meter mirror keeps the image centered in the focal plane.

So far the OWL team has only preliminary ideas for adaptive optics. They are basing their hopes on a demonstration in the summer of 2004 of ESO's Multi-Conjugate Adaptive Optics system (MCAO) on one of the VLT units. If the ESO design could be extended to OWL, it would deliver diffraction-limited resolution over a field of view of 2 arcminutes. That translates to 4 milli-arcseconds at a wavelength of 2 microns. But there is a long road ahead to achieving this goal.

The team has redesigned the telescope's mechanical structure several times in order to reduce the mass of steel that would be required. In the current alt-azimuth design, the ratio of weights of steel to glass has been reduced to 7, as compared with a conventional ratio of 20. Nevertheless, the total weight of the telescope is 12,500 tons, about twice the weight of the Eiffel Tower.

The primary mirror and its correcting optics turn in altitude within a steel cradle 60 meters in diameter (see fig. 11.12). The cradle rests on an azimuth ring 170 meters in diameter, and the whole structure rotates in azimuth on 250 sets of wheels (bogies). Each bogie has four spherical wheels that are independently driven. This arrangement enables the telescope to be pointed in azimuth to within 3 arcminutes of the desired direction, or one-tenth of the Moon's angular diameter.

Although the telescope structure is partly below ground, its top rises to a height of 95 meters. How do you shelter such a huge and delicate machine from the elements? The team proposes to operate the telescope in the open air, outside an enclosure. When bad weather threatens, the enclosure could be rolled over the telescope and sealed. Observations might have to be limited to periods when the wind is light, however. A strong wind blowing toward the telescope could disturb the shape of the primary, while a wind blowing from the rear could generate undesirable turbulence in the light path. The primary and secondary mirrors are under active control to correct for such disturbances, but their effectiveness has yet to be tested.

The actual size of an enclosure is not a problem; hangars for dirigibles as large as 360 meters long have been built. Nevertheless, an enclosure for the OWL could cost EUR 70 million. The full cost of an OWL is approximately EUR 1 billion.

ESO is pursuing this dazzling concept with the idea of leapfrogging an intermediate step to, say, a 30-meter. It is a bold venture, with high risks but huge benefits. Even without adaptive optics the OWL could carry out pioneering science with its powerful spectrographs. If all goes well, a first-generation adaptive system could be in operation by 2017 and a complete system by 2021. Time will tell whether ESO has made the right decision.

THE TINY LITTLE RADIO WAVES

In contrast to the TMT and the OWL, the Large Atacama Telescope (LAT) would be a bargain at a mere $60 million. This 25-meter submillimeter telescope would be located near the Atacama Large Millimeter Array (ALMA) in the Chilean Atacama desert, at an altitude of 5,000 meters. Although the ALMA would be more sensitive and have higher angular resolution, the LAT would have a much larger field of view and would excel in mapping the large-scale structure of the intergalactic medium.

In February 2004 Cornell University and Caltech signed a memorandum of understanding to share the costs of the telescope design. The design will take two years and cost $2 million. Cornell has already received $250,000 toward its share from Fred Young, a Cornell alumnus. An agreement on construction and operational costs may follow later.

LOOKING FOR THE PERFECT SITE

Building an ELT is a slow, expensive process. After the design is complete and has passed reviews, after the technology has been developed, after the necessary funds have been accumulated—only then can construction begin. A decade or more will pass before any of these giants sees first light. The delay may be frustrating, but the sponsors of these telescopes are putting the time to good use by evaluating potential sites. A site survey should last a minimum of three to five years, given the variability of some factors.

Some groups have already chosen a general area and only need to refine their search. The Large Atacama Telescope team has decided on a specific site in Chile. The Giant Magellan Telescope would almost certainly be located in Chile, possibly near the other Magellan telescopes at Cerro Paranal. And the Euro50 would be built somewhere near the other telescopes on La Palma.

The OWL and Thirty-Meter Telescope teams, on the other hand, are still searching for the best sites. They are sharing their data, with a possible view toward sharing the costs of developing a new site. ESO would certainly prefer Chile for the OWL. On the other hand, the TMT group is undecided whether to build in Chile or somewhere in the North American Southwest. Although Chile offers attractive sites as well as access to the relatively unexplored southern sky, political factors and donor preferences may also enter into the decision.

Everyone agrees on the criteria for choosing a site, although priorities differ somewhat. Clear nights and good seeing probably top the list. With all the emphasis on infrared observations, low humidity at the site is essential. A dark sky and favorable wind conditions are also important. Ultimately the costs of developing a remote site enter into the calculation.

Site testing has come a long way since the 1970s, when astronomers were prospecting in Mauna Kea and the Canary Islands. More sensitive instruments, automatic data recording, and better methods of analysis have changed site selection from an art to a science. The survey undertaken by the TMT group is typical of modern techniques.

Their first step was to examine five years of satellite data for cloud cover and water vapor. These data have improved significantly since the 1970s. Observations at 10 microns reveal the presence of low-level clouds, while those at 6 microns detect water vapor and cirrus clouds. Three prime locations in the Northern Hemisphere emerged from this study: southern California, western Arizona, and northern Baja California, Mexico. The survey team also reviewed previous surveys by AURA in Hawaii and the southwestern United States and by ESO in Chile.

With these results in hand the team is planning to set up instruments at a small number of sites, to measure the seeing and monitor the weather. One cannot overstate the importance of good seeing for a telescope 30 meters or larger. Part of the justification for building such a telescope is the science that could be done near its diffraction limit of resolution. Adaptive optics systems can deliver such performance, but only if the unaided seeing at the site is quite good to begin with. And the better the seeing is, the simpler and cheaper an adaptive optics system can be.

As we learned in earlier chapters, the profile of turbulence through the atmosphere determines the quality of a telescope's images. The larger the turbulence cells and the more slowly they change, the sharper a telescope's images. For a first look at the effects of turbulence, the TMT team will measure the size of the dominant turbulent air cell (or, more precisely, the Fried parameter, r_0) at different wavelengths.

They have built several copies of the Differential Image Motion Monitor, or DIMM, that ESO used to find a site for the VLT. It is a refinement of the double-beam telescopes used in older surveys (see chapter 4). The basic idea is to place a mask over the mirror of a small telescope. Two small holes in the mask, which are separated by the diameter of the mirror, form two images side by side in the focal plane. The relative motion of the images, recorded over a period of several hours, indicates the sizes of the turbulent air cells above the ground. In 1990 M. Sarazin and François Roddier worked out the detailed theory to extract the Fried parameter from the data. In practice the DIMM is mounted on a platform 5 meters high to avoid the turbulent ground layer that would lie below a large telescope.

A DIMM yields one important measure of the seeing. With more detailed information on the height variation of turbulence, one could calculate all the seeing characteristics for a large telescope and its adaptive optics system. To obtain such data, the TMT team will use several clever devices.

Jean Vernin and François Roddier invented Scintillation Detection and Ranging (SCIDAR) in 1973. The technique is based on the idea that the apparent displacement of a star (*twinkling*, or scintillation) by a turbulent layer depends on the height of the layer. In general, higher layers are more effective than lower layers. Vernin and Roddier showed how to analyze a sequence of images of single or double stars to obtain the height distributions of turbulence and the wind speed. Their method has already been applied in many site surveys, including those for the Very Large Telescope, the Nordic Optical Telescope, and the Canada-France-Hawaii Telescope.

More recently, Victor Kornilov and his colleagues at Moscow University's Sternberg Astronomical Institute invented the Multi-Aperture Scintillation Sensor (MASS). Like SCIDAR, it uses the scintillation of a star to determine the strength of turbulence at different heights in the atmosphere. Unlike SCIDAR, however, it depends on the variations of brightness caused by turbulence. In its present design MASS can sample the turbulence at six discrete heights (note 11.5). Moreover it has the advantage that it will operate automatically for long periods at a remote site.

The first several hundred meters of air above a site are the most critical for seeing. To sample this layer at closer intervals than is possible with MASS, the TMT team is also considering Sound Detection and Ranging, or SODAR. This system sends a pulse of sound upward and times the back-scattered echoes. The arrival times indicate the presence of turbulent layers at different heights. With sufficient signal strength, one can determine both the turbulence and the wind speeds aloft.

To augment SODAR, the team plans to use another well-tested technique. They will erect a tall mast that is equipped with sensitive thermometers *(microthermal sensors)* and wind speed meters *(anemometers)*. Turbulent air fluctuates in temperature by a few thousandths of a degree on time scales of a few milliseconds. These variations yield useful information on the physical scales of the turbulence.

Along with this high-tech equipment, the TMT team will use standard automatic weather stations and cloud-cover monitors at each prospective site. Once candidate sites are chosen, the team will settle down to a long campaign to select the very best one.

The next generation of extremely large telescopes will be awesome machines, the products of high technology and soaring imagination. The urge to build them is motivated by the anticipated needs of the science, by developments in technology, and in part by the aspirations of a few key individuals. National pride, professional competition, and the excitement of the general public are involved as well. And yet each ELT will cost hundreds of millions of dollars. How many can the nations of the world afford? How do the needs of astronomy compare with those of other branches of physics, biology, and medicine? The public and their representatives in government and at the great private foundations will have to decide.

I have no doubt that one way or the other, one or more of these giants will be built, with private or public money or both. The science they offer is just too exciting to ignore.

THE FUTURE IN SPACE

THERE MAY BE a limit to the size of a ground-based telescope, although judging from the proposal to build the Overwhelmingly Large Telescope, we haven't reached it yet. In space, telescopes are not getting bigger so much as they are getting smarter. There is a practical limit to the size of a mirror one can launch into orbit, set by the size of the available launch vehicles. Perhaps after the International Space Station is complete, we'll attempt to assemble something really huge in orbit or on the Moon, but for the foreseeable future, space telescopes must rely on mirrors that are modest by standards on the ground (but see note 12.1). They will, however, employ novel focal plane instruments and clever detection methods to push the envelope.

In this chapter we'll discuss an assortment of space-based optical/infrared observatories that will be launched over the next decade. Several of them, like the James Webb Space Telescope, will be simply larger and better-equipped versions of earlier missions. Others, like the Kepler Mission and the Terrestrial Planet Finder, will employ novel detection schemes. Some, like the European Space Agency's Darwin project, are quite breathtaking in their audacity.

Most of the observatories divide into two camps according to their main scientific program. One group will focus on the evolution of the early universe, the other on a search for extraterrestrial planets. A few mavericks refuse to be categorized so easily.

These observatories wouldn't be built if they didn't have an overwhelming advantage in some respect over their ground-based brethren. Their most obvious advantage is access to parts of the spectrum that are difficult or impossible to reach from the ground. Hence the drive to build satellites to observe in the far infrared. Other advantages are not so obvious, as we shall see.

Let's begin with the latest infrared observatory to reach orbit, the Spitzer Space Telescope. It has had a long and troubled adolescence.

SPITZER, A LONG TIME IN COMING

As we learned in chapter 11, every ten years the U.S. National Academy of Sciences convenes a committee to survey trends in astronomy and to recommend funding priorities for NASA and the National Science Foundation. Way back in 1979 a committee, chaired by Princeton astrophysicist George Field, recommended that NASA should build the Shuttle Infrared Telescope Facility (SIRTF). This actively cooled telescope would be carried aboard the space shuttle to observe the sky at wavelengths beyond 10 microns. The committee predicted that such observations would open a broad frontier in astronomy.

NASA engineers set about designing the spacecraft, but first, in partnership with the Dutch and the British, it launched the Infrared Astronomical Satellite (IRAS) into a polar orbit. IRAS was the first space telescope to be cooled with liquid helium. At a temperature of 4 kelvin it was able to observe in the far infrared. Over a period of ten months in 1983 IRAS scanned the entire sky in four broad bands centered at 12, 25, 60, and 100 microns. It discovered a gold mine.

Over 250,000 sources were pinpointed with an accuracy of about 20 arcseconds, and spectra were obtained for a special 5,000. Among IRAS's major discoveries was an accretion disk around the star Vega; six dusty comets; starburst galaxies, where star formation was under way at a frantic rate; strong emission from colliding galaxies; and warm interstellar dust that pervades space in all directions. In addition, the opaque core of the Milky Way was penetrated for the first time (fig. 12.1). IRAS was a stunning success that unleashed a flood of scientific papers. Astronomers everywhere were eager to follow up its discoveries with SIRTF.

NASA engineers initially designed the SIRTF to operate from the cargo bay of a space shuttle. But a small infrared experiment flown aboard the shuttle in 1985 revealed that the shuttle environment was badly contaminated with heat and vapors. The shuttle would be totally unsuited for observations at long infrared wavelengths. Therefore, SIRTF would have to be redesigned as a free-

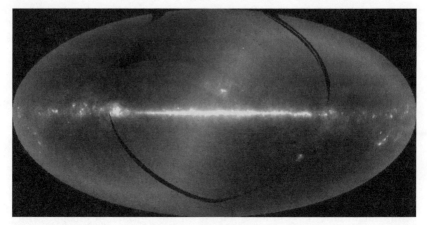

Fig. 12.1. The whole sky as imaged at 12, 60, and 100 microns by the Infrared Astronomical Satellite. In this composite picture blue indicates hotter regions, red cooler regions (see color gallery). The horizontal band is the plane of the Milky Way. The dark curved feature is produced by warm dust in the plane of the solar system.

flyer. Its name was changed to the *Space* Infrared Telescope Facility, and preparations were made for a flight that could last several years. Meanwhile, NASA was engaged in building and flying the Hubble Space Telescope.

The SIRTF project moved along slowly. By 1990 a decade had passed since the Field committee's report. The National Academy of Sciences convened a new committee, chaired by John Bahcall of the Institute for Advanced Study at Princeton. It endorsed once again the concept of a free-flying infrared telescope that would be cooled by liquid helium. It recommended that SIRTF should be equipped to observe the full range from 1 to 1,000 microns. A long list of possible research areas could then be addressed, including such topics as the formation of stars and planets, the origin of quasars, the distribution of galaxies, and the formation and evolution of galaxies. NASA needed no urging, but unfortunately the agency encountered a series of budget constraints over the following years. SIRTF was "descoped" several times and reduced in cost from $2 billion to $500 million.

Meanwhile, the Europeans were pushing ahead. In 1995 the European Space Agency launched the Infrared Space Observatory (ISO), a 0.6-meter cryogenic telescope equipped with the first two-dimensional infrared detectors. During a flight of thirty months ISO examined nearly thirty thousand individual objects at wavelengths between 2.4 and 240 microns. It detected water molecules everywhere in our galaxy and discovered many new interstellar molecules. It explored active galactic nuclei; the birth of stars in nearby dark clouds (fig. 12.2) as well as in distant galaxies; planetary atmospheres; dusty comets; and much more. It was a brilliant success. American astronomers were eager to catch up with the Europeans.

Progress in building SIRTF accelerated in the 1990s. NASA engineers and their contracting partners came up with several bright ideas in technology and orbital design. First, all of the telescope parts, including the mirror, were built of beryllium. Beryllium is a gray, brittle metal, very difficult to machine and polish to a high shine. But it is light, only two-thirds as dense as aluminum. Beryllium is in fact the fourth lightest element, heavier only than hydrogen, helium, and lithium. The lightweight mirror would help to reduce the overall weight of the spacecraft.

Second, SIRTF would be launched into a special orbit around the Sun. It would trail the Earth in its orbit, slowly drifting away at a rate of 15 million km per year. As it left the vicinity of the warm Earth, the telescope would cool

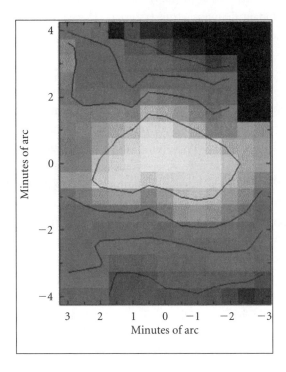

Fig. 12.2. The Infrared Space Observatory captured the first image of a dense core of gas molecules and dust in which a star will form. This picture was obtained at a wavelength of 160 microns.

down naturally to the ambient temperature of 30 kelvin. The load on the on-board cooling system would therefore decline, and the supply of helium could be conserved.

Third, the spacecraft was built with a warm zone, where the electronics work, and a separate cryogenic zone, which contains the telescope mirror and the science instruments.

After a long evolution, SIRTF was launched on August 25, 2003, for a nominal thirty-month mission. It was the last of the four "Great Observatories" that include the Chandra X-Ray Observatory, the Compton Gamma Ray Observatory, and the Hubble Space Telescope. Like the others, it was renamed in honor of a pioneering astrophysicist, Lyman Spitzer of Princeton University (see chapter 5).

The Spitzer Space Telescope (SST) has four main science objectives: to observe galaxy formation in the early universe; to search for brown dwarfs and massive planets that might qualify as dark matter; to study star formation in luminous galaxies; and to search for examples of planetary formation.

The telescope is a Ritchey-Chrétien with a 0.85-meter mirror, diffraction-limited at all wavelengths longer than 6 microns (fig. 12.3). Three science instruments, all cooled to 6 kelvin with liquid helium, will carry out the science programs. Giovanni Fazio, a researcher at the Smithsonian Astrophysical Observatory, led the consortium that built the Infrared Array Camera (IRAC). It takes pictures simultaneously at four wavelengths between 3.6 and 8 microns, using square arrays of 256 pixels on a side. James Houck, an astrophysicist at Cornell University, is project scientist for the Infrared Spectrograph (IRS). This instrument has four channels between 5.3 and 40 microns and is provided with 128×128 pixel arrays. Finally, George Rieke (University of Arizona) is the principal investigator for the Multiband Imaging Photometer. Its 32×32 pixel arrays capture images at three wavelengths, 24, 70, and 160 microns.

One unfortunate aspect of the long delay in finishing the SST is the relatively small size of its infrared detectors. The instruments were designed and built long before detectors became available with up to a million pixels. So to a certain extent the designs were frozen. This is a disadvantage many space telescopes share. They take so long from conception to launch that their technology is often dated by the time they are deployed. The Hubble's instruments could be updated because of its nearby orbit, but the James Webb Space Telescope, for example, will be too far away to service.

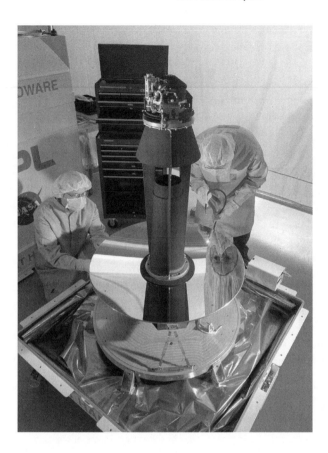

Fig. 12.3. Technicians working on the beryllium mirror for the Spitzer Space Telescope

The SST is a national facility, open to any qualified scientist. Observing time is awarded by the Spitzer Science Center, which solicits proposals, has them reviewed by experts, and selects the best. To inaugurate the SST, a so-called Legacy Science Program of six observing programs has been organized. In addition to achieving particular scientific goals, these programs are intended to establish a useful database for future researchers. Here is a sample of these early programs:

- Edward Churchill from the University of Wisconsin leads a large and varied academic team. They are mapping the inner parts of the disk of the Milky Way at a resolution of only 1 arcsecond, in order to pick out discrete infrared sources that are heavily obscured by dust. They are using the Infrared Array Camera to obtain images at four wavelengths between 3 and 8 microns.

- We learned in previous chapters about the Deep Fields that the Hubble Space Telescope viewed at optical wavelengths. Now Mark Dickinson (Space Science Institute) and his colleagues will remap the same fields with the Array Camera at 3 to 8 microns and with the Infrared Spectrograph at 24 microns. They hope to study the assembly of normal galaxies in the early universe at redshifts of $z = 1$ to 6 and the energy generation due to star formation.

- Neal Evans (University of Texas) and his crew have an ambitious goal. They will attempt to observe the complete sequence of the formation of a star from a molecular cloud to the evolution of a planetary disk. They have compiled an extensive list of targets: 150 known clouds and at least 190 candidate planetary systems. The Array Camera and the Infrared Spectrograph are their main tools.

- The SIRTF Nearby Galaxies Survey (SINGS) is the name of the program headed by Robert Kennicutt Jr. (University of Arizona). He and his team have selected seventy-five galaxies in which they plan to relate the properties of the interstellar gas to the process of star formation. Utilizing optical as well as infrared observations of these galaxies, they hope to disentangle the dust spectra from the stellar spectra. Then they would be able to understand how the molecular and dust composition of the interstellar medium affects the rates of star formation.

- Carol Lonsdale (Caltech) leads a huge cast of characters in a program called the SIRTF Wide-Area Infrared Extragalactic Survey (SWIRE). These folks will seek out galaxies at redshifts of $z = 1$ to 3 in seven separate regions of the sky. Using images and spectra at all seven wavelengths between 3 and 70 microns, they will try to reconstruct the evolution of galaxies and its dependence on the local environment. The growth of active nuclei in galaxies is especially interesting to this group. They anticipate harvesting a huge sample of over a million galaxies in their quest.

- Finally, a large team led by Michael Meyer, of the University of Arizona, has proposed to study the formation and evolution of planetary systems. They have selected two groups of Sun-like stars, each containing 150 members that differ substantially in age. The distribution of dust and the motions of gases around these stars will be determined by means of a combination of observations and model calculations. The team wants to determine how planet formation depends on the characteristics of the local environment of a star. Does the formation of planets require special circumstances? Is our solar system special in this sense?

These Legacy programs are just getting started, so it's too soon to expect detailed results. But the Spitzer Space Telescope has already transmitted some exciting images.

Stars form in cold interstellar clouds when some agent triggers an instability. A supernova is one type of agent. When a dying star explodes, it ejects a high-speed shell of gas that sweeps up interstellar gas like a snowplow. At the front edge of the shell, the gas is compressed, its density rises, and a spontaneous collapse occurs. A star is born.

In March 2004 the SST caught a graphic example of this type of event in the Large Magellanic Cloud, one of the Milky Way's satellite galaxies. Figure 12.4 shows the slow-motion birth of a cluster of stars in a particular nebula, Henize 206. The SST was able to pierce the dense clouds of dust surrounding this star nursery by looking in the far infrared.

Fig. 12.4. One of the first images from the Spitzer Space Telescope shows this stellar nursery in the Large Magellanic Cloud. An expanding supernova remnant *(upper part of the image)* has triggered the birth of stars in this nebula, Henize 206.

Then in May 2004 Edward Churchwell (University of Wisconsin) and his team announced the discovery of the youngest planet found so far, an infant less than a million years old. The baby planet nestles in a warm cradle of dust around a young star in the constellation Taurus. The team also discovered a region in Centaurus that contains about three hundred young stars. Churchwell said that the formation of stars and planets in our galaxy seems far more common than he and his team had expected.

After a long delay, the Spitzer Space Telescope is off to a good start.

EUROPE'S BIG GUN

The European Space Agency is building what will be the largest infrared telescope in space when it is launched in 2007. Herschel, formerly called FIRST, will have a 3.5-meter telescope. It will be the first satellite able to observe very faint objects over the whole range from the far infrared to the submillimeter. At wavelengths between 60 and 670 microns, it will be able to sample some of the coolest objects in space and also to penetrate the densest veils of dust.

Herschel's main objective is to explore the early universe, when galaxies were beginning to form out of the cosmic soup. ESA scientists want to learn how these first galaxies were assembled, when the first stars lit up, and how clusters of galaxies evolved. They also want to see how stars in our own galaxy form in dense molecular clouds, and how circumstellar disks evolve into proto-planets. And they want to investigate the molecular chemistry of interstellar gases.

To carry out this scientific program, three instruments are planned for spectroscopy, photometry, and imaging. A consortium of European and American astronomers, led by a Dutch team, is building the Heterodyne Instrument for the Far Infrared (HIFI). In this state-of-the-art instrument, the light from a cosmic source is mixed with a locally generated signal (note 12.2). It is capable of very high resolution spectroscopy in seven bands ranging from 160 to 600 microns.

The Photoconductor Array Camera and Spectrometer (PACS) is the brainchild of Albrecht Poglitsch and his team at the Max Planck Institute for Extraterrestrial Physics. This dual-purpose instrument produces low-resolution spectra of the sources within a 50-arcsecond field of view. In a different operating mode, the PACS measures the brightness of the sources within a field of 3 arcminutes, in three bands between 60 and 210 microns.

Finally, the Spectral and Photometric Imaging Receiver (SPIRE) is being built by a consortium of five European nations and the United States. This instrument is designed to detect large populations of faint distant galaxies and then to capture their spectra at low-wavelength resolution. To find the galaxies, it will record images within a huge 4 × 8 arcminute field of view, in three bands from 250 to 520 microns. Then it will use its Fourier Spectrometer, an elegant device for high-resolution spectroscopy, on selected targets.

Herschel's components will be separated so as to work at three very different temperatures. The instruments will be cooled with liquid helium in a special container to about 2 degrees kelvin above absolute zero. Meanwhile, the telescope mirror will be allowed to chill down passively in the frigid dark to an estimated 80 kelvin, and the vital electronics will work at room temperature in a separate enclosure.

Herschel's telescope is a classical Cassegrain with high-tech improvements. Its mirror is composed of twelve segments of silicon carbide, which is light in weight, has excellent thermal properties, and can be polished to the required accuracy. The faceplate is only 2 mm thick but is supported with spokes of silicon carbide. The mirror is being built to survive the violent shaking during launch and the huge change of temperature as it cools down in space.

ESA plans to launch Herschel in 2007 from its (French) Guiana Space Center to the L2 Lagrangian point, 1.5 million km from Earth. The observatory will be stocked with sufficient helium to last at least three years. Herschel will be available to astronomers everywhere on a competitive basis. An ESA committee will allocate two-thirds of the total observing time to the best proposals. The line is forming already. We can expect to hear some wonderful news from Herschel during, and long after, its flight.

JAMES WEBB, BIGGER BUT LATER

If all goes well, NASA will launch the James Webb Space Telescope (JWST) in 2012, as the successor to the Hubble Space Telescope and the Spitzer Space Telescope. With its 6.5-meter mirror, it will collect sixty times as much infrared light as the Spitzer Space Telescope and three times as much as Herschel. Like

Herschel, its main mission will be to observe the universe at an age of less than a billion years after the Big Bang. This was the era in which the first galaxies and the first stars were formed out of the primordial gases. These primitive objects lie at the extreme limits of the observable universe, at redshifts of $z = 6$ to 12 and more.

The JWST will be equipped with cryogenically cooled detectors for the near and middle infrared, from 1 to 28 microns. In this respect, the JWST and Herschel are complementary; Herschel covers the spectrum beyond 60 microns, while the JWST observes short of 30 microns.

The origin and evolution of the first galaxies is only one of several interesting topics that JWST will pursue. Our Milky Way galaxy and its neighbors will also receive some intense scrutiny. Astronomers want to understand when and how the central bulge of the galaxy formed, and how the galaxy's chemical composition varies among its different parts. How old are the oldest stars? When and how did the Large and Small Magellanic Clouds form?

Then there is the question of precisely how stars form. How do the local properties of the interstellar gas and dust influence the rates of star formation? What determines the maximum mass of a star? How do circumstellar disks arise, and how are proto-planets formed? These and other questions like them will occupy astronomers for years to come.

Although dark matter constitutes 90 percent of the mass of the universe, we still have no agreement on what it is. Heavy neutrinos and other exotic elementary particles have been pretty well ruled out. Brown dwarfs, stars of very low mass and brightness, may contribute to the dark matter, but we lack enough information about their density in space to decide. As a first step, astronomers will use the JWST to observe the gravitational lensing (see note 11.1) of distant galaxies, and so determine the large-scale distribution of dark matter.

We touched on the design of the JWST earlier, in connection with the development of the Extremely Large Telescopes. Just to remind you, the JWST's mirror is constructed of eighteen segments that will unfold in space like a giant flower. If it were made of glass, like the 6.5-meter mirror at Las Campanas, the mirror would weigh more than 10 tons, an awesome load to lift into space. Therefore NASA's contractor has chosen lightweight beryllium as the mirror's material. So although the JWST's mirror is 2.5 times as large as the Hubble's, it weighs only one-third as much.

The JWST will orbit the Sun at the L2 Lagrangian point, where it will have a continuous view of the sky. In outer space the telescope will chill down to an estimated temperature of about 40 kelvin, cold enough to prevent the telescope's heat from swamping the onboard detectors.

The JWST will have three times the collecting area of the Herschel. That should give it an advantage in exploring the formation and evolution of galaxies and stars in the early universe, at redshifts of $z = 6$ and higher. It will be able not only to observe fainter objects but also to resolve finer details. It will be the premier instrument for far-infrared research, but it may lag Herschel by four or five years. If present schedules at both agencies hold, the Europeans will be well positioned to skim much of the cream in this exciting new field.

The real competition may come down to the focal plane instruments of the two space observatories. In November 2001 NASA issued a call for proposals from university and industrial scientists to build the instruments for the JWST and received some excellent responses. Three cryogenically cooled instruments were selected: a near-infrared camera (NIRCam), a near-infrared spectrograph (NIRSpec), and a mid-infrared instrument (MIRI).

Marcia Rieke, a well-known expert in space infrared and planetary astronomy, heads the team for the NIRCam at the University of Arizona. This instrument will take pictures at wavelengths between 0.6 and 5 microns. Each of its state-of-the-art infrared detectors has four million pixels. So, for example, an image could have a resolution of 0.12 arcsecond over a field of 2 arcminutes, at a wavelength of 4 microns. A whole cluster of galaxies could be imaged in sharp detail. This clever device is also designed to block the light of a star, so as to enable a researcher to look for its planets.

George Rieke, Marcia's husband and also a world-class infrared astronomer, leads the international science team for the mid-infrared instrument, MIRI. This combination of imager and spectrograph covers the critical wavelength range between 5 and 27 microns. Combined with the enormous light-collecting power of the JWST, MIRI will offer a gain in sensitivity of a thousand over existing ground-based telescopes. It holds the key to the exploration of the early universe.

MIRI's images will be diffraction-limited at 5 microns over a field of view of 1.5 arcminutes. The imager will also offer low-resolution spectra of all the objects within this wide field. But the high spectral resolution required for detailed investigations will be provided by a separate spectrograph. This instrument will have a very limited field of view, perhaps as small as 2 arcseconds; that should be sufficient to target a single galaxy.

The European Space Agency and NASA are collaborating in the construction of a near-infrared spectrograph (NIRSpec). ESA will furnish the main parts of the instrument while NASA will provide the 2,000 × 2,000 pixel arrays. NIRSpec will operate between 0.6 and 5 microns. It will be able to obtain the spectra of over one hundred objects simultaneously, in a field of view of 3 × 3 arcminutes. It will be ideal for rapid surveys of galaxy clustering, young stellar clusters, and active galactic nuclei. A Canadian team is providing a fine guidance system for the JWST. It includes a tunable filter for the range between 1 and 4.8 microns.

The teams that build instruments for NASA's satellites used to have exclusive use of them during the first year of operations. That policy was eventually modified. Now the teams have to compete for time with "guest investigators," who may propose observing programs that utilize all the instruments on board. But if you have built an instrument and know all about its quirks, you do have an advantage. So the general practice now is for the builders to collaborate with the guests. Hopefully, everybody wins. We shall see.

JAPAN'S ENTRY

Japanese astronomers have pioneered in x-ray and gamma-ray astronomy, but their efforts in space infrared astronomy have been more limited. In 1995 they launched their first infrared observatory. The Infrared Telescope in Space (IRTS) was cryogenically cooled to 2 kelvin. It surveyed approximately 7 percent of the sky for twenty-eight days, at wavelengths between 1 and 1,000 microns. Its 15-cm mirror directed light to four focal plane instruments: two grating spectrometers for the ranges of 1 to 4 and 5 to 12 microns; an instrument to monitor spectral lines at 63 and 158 microns; and a broad-band spectrometer for the range 150 to 700 microns.

Among the more interesting results to emerge from the flight was the discovery of a variation of the temperature of interstellar dust above the plane of the Milky Way. Takanori Hirao and his colleagues observed the emission in four bands between 150 and 700. They determined that the temperature of the dust was a "warm" 20 kelvin but may decrease to 4 kelvin away from the galactic plane.

In Japan, large astronomical projects are allocated among several research groups in a fairly reliable cycle. Infrared astronomers at the Institute of Space and Aeronautical Science will get their chance again in 2005. They have designed ASTRO-F (formerly called the Infrared Imaging Surveyor, IRIS) as a liquid-helium-cooled telescope. It is intended to map the infrared sky with a resolution and sensitivity at least ten times higher than achieved by IRAS.

Hiroshi Shibai and his colleagues expect to see as many as ten million galaxies. Many of these will be luminous starburst galaxies at redshifts greater than $z = 1$. In addition, ASTRO-F will search for brown dwarfs and protoplanetary objects. The data it will acquire will be useful in studying the formation of galaxies, stars, and planets.

ASTRO-F is a Ritchey-Chrétien telescope with a 0.67-meter mirror, cooled to 6 kelvin. The mirror is made of silicon carbide and is coated with gold to enhance its reflectivity at infrared wavelengths. Two focal plane instruments have been built. The Far Infrared Surveyor records the light in four bands between 50 and 200 microns. The Infrared Camera actually contains three cameras that cover the range from 2 to 26 microns. They have the great advantage of wide fields of view, up to 10 arcminutes.

The Japanese team plans to launch ASTRO-F in a Sun-synchronous orbit that passes over the north and south poles of the Earth, at the twilight zone. To scan the sky, the spacecraft will spin slowly in its orbit. The telescope can also be pointed in specific directions to obtain low-resolution images and spectra. ASTRO-F passed its critical shake tests in December 2004. If all goes well, ASTRO-F will be launched in 2005 for a flight of 550 days, limited only by the supply of helium.

What's next for Japanese astronomers? Takao Nakagawa and his colleagues at the institute are already designing the next generation of infrared observatory. The Space Infrared Telescope for Cosmology and Astrophysics (SPICA) would be a 3.5-meter, cryogenically cooled to operate in the middle and far infrared. To reduce the weight of the spacecraft the telescope would be launched at room temperature and allowed to cool in space, like the Spitzer Space Telescope. A small onboard cryostat would then chill the instruments to 4.5 kelvin.

The SPICA team has the same scientific goals as other teams in Europe and the United States, namely, tracing the history of galaxy formation in the early universe, the birth and evolution of active galactic nuclei, and the formation of stars and planets. SPICA could be launched to orbit around the L2 Lagrangian point as early as 2010.

THE MILKY WAY IN THREE DIMENSIONS

As the saying goes, "There's no place like home." While many astronomers are chasing galaxies at the far reaches of the universe, a few conservatives are focusing on our own neighborhood. The European Space Agency is building GAIA, an astrometric satellite that will create a three-dimensional map of a billion stars in the Milky Way. That's only 1 percent of the one hundred billion our galaxy is estimated to contain, but enough to achieve several important scientific goals.

GAIA is an acronym for Global Astrometric Interferometer for Astrophysics. The satellite is appropriately named after the Greek goddess of the Earth and nature. GAIA will be the successor to the first astrometric satellite, Hipparcos, which ESA launched in 1989. Hipparcos is an acronym for High Precision Parallax Collecting Satellite (note 12.3). This pioneering satellite determined the positions on the sky of 120,000 stars with a precision of 2 milli-arcseconds and a million more stars at a precision of 20 milli-arcseconds. It also obtained their brightness in two colors.

By measuring the shift in the apparent position of each star as the Earth orbited the Sun (the *parallax*), astronomers could determine the star's distance (note 12.4). In this way, Hipparcos improved and extended the distance scale to the nearest stars, a vital step in calibrating distances of much more distant objects. For example, Hipparcos revised the distance to the nearest Cepheid variables, which led to an increase of 10 percent in the estimated size and age of the observable universe.

Data from Hipparcos also revealed that the disk of the Milky Way is warped, like the brim of a hat. And by repeating the positional measurements over the life of the mission, Hipparcos pinned down the angular speed *(proper motion)* of hundreds of thousands of stars. That was useful in distinguishing stars of different ages.

Hipparcos was a triumph, but GAIA will do better. After it is launched in 2010, GAIA will measure the position, brightness, and colors of a billion stars.

During its five-year flight, the satellite will repeat these measurements one hundred times. This mountain of data will fix the stars' distances, proper motions, and spectral types. Not only will GAIA measure ten thousand times as many stars as Hipparcos, it will fix their positions a hundred times more accurately.

You may wonder how a satellite whirling around in orbit can measure the positions of stars with a precision of 10 *micro*-arcseconds. (That's the width of a human hair at a distance of 400 km.) The basic idea is to take simultaneous overlapping pictures of the sky with two cameras that are almost completely free of distortions. Each image contains some stars that also appear in the neighboring images, so that the stellar positions can be tied into a global grid. Then, the parallax method can be applied to the seasonal changes in the apparent positions of the stars.

Like Hipparcos, GAIA will have two telescopes for astrometry and a third for low-resolution spectroscopy. Each astrometric telescope has a rectangular mirror measuring 1.4 by 0.5 meters. The telescopes will point in directions that differ by a fixed angle of 106 degrees. As the satellite spins, the two telescopes will sweep across the sky and form images side by side on an enormous array of 180 CCD detectors. (Each detector will have 8 million pixels.) During a rotation, each telescope scans the same patch of sky, which helps to nail down the relative angular positions of the thousands of stars in each image. The digital data are transmitted in a continuous stream, hour by hour, year by year. In time, GAIA will enable astronomers to create a three-dimensional map of our Milky Way galaxy.

Michael Perryman, an astronomer at the European Space Research and Technology Center, was the project scientist for Hipparcos and now plays the same role for GAIA. He has high hopes for the science that GAIA can produce. One result would be a history of our galaxy. Each star that formed in the galaxy retains the motion and the chemical composition it was born with. These quantities are indicators of the age of the star. By determining the location, speeds, brightness, and colors of stars in different parts of the galaxy (the disk, the halo, the central bulge, and the spiral arms), GAIA will enable astronomers to recover the history of our galaxy. How and when did the globular clusters form, far from the central disk of the galaxy? When did the disk and arms appear? How did the disk spin up as the galaxy collapsed upon itself? For the first time we may find answers to such questions.

SEVERAL WAYS TO SKIN A CAT

Most people who have access to television or to the Internet are fascinated by the possibility that life exists outside our own small solar system. Carl Sagan and Frank Drake, among others, have offered persuasive reasons why life forms should or could be common throughout the universe, and the public is waiting patiently for a discovery. The probability of finding a habitable planet around any particular star is minuscule, but with so many billions to search, the prospects of finding one somewhere are reasonable. Certainly NASA scientists think so, and the agency has undertaken a massive effort to search for them, in a program called Origins (see note 11.3).

Extrasolar planets became a hot topic in 1995, when Swiss astronomers Michel Mayor and Didier Queloz discovered a Jupiter-sized companion to the star 51 Pegasi. They deduced the presence of an invisible planet from periodic variations in the star's radial velocity. Geoffrey Marcy (University of California, Berkeley) and Paul Butler (University of Maryland) used the same method to discover many of the 122 extrasolar planets known today. Most of these planets are Jupiter-sized giants that approach very close to their host stars. Their surface temperatures would generally be too high to allow life forms to survive. In fact, astronomers are puzzled as to how such giants could form or survive so close to their host stars.

Although radial velocity surveys with highly stable spectrographs have been the most productive means of finding planets so far, astrometry offers a different approach. Periodic shifts in the *position* of a star would reveal the presence of an unseen companion. This is the technique that the GAIA team plans to use. And as we have seen, ground-based interferometric arrays such as the Keck Interferometer and the Navy Prototype Optical Interferometer (NPOI) have been built with this idea in mind. A special technique, *nulling interferometry,* can also be used to block out the light from a candidate star and obtain an actual image of a planet. No ground-based interferometer has discovered a planet to date, but the prospects with a satellite are much brighter.

Yet another scheme for detecting planets is based on precision photometry. In the most favorable situations, we could see a planet crossing the disk of a star (*transiting,* as the astronomers say) if we happened to lie in the plane of the planet's orbit. Because a planet is opaque, the amount of light we would receive from an unresolved star would therefore decrease. We actually see this

Fig. 12.5. On May 7, 2003, the Solar Oscillation and Heliospheric Observatory followed the transit of Mercury across the disk of the Sun. This composite image shows the tiny dark planet at many instants during its five-hour crossing. During its transit Mercury decreased the amount of sunlight we received by a mere 0.004 percent.

dimming effect when Mercury or Venus transits across the face of the Sun (fig. 12.5). Venus performed this sweep for the first time since 1882 on June 6, 2004. The Sun's brightness, as seen at Earth, decreased by a mere one-tenth of 1 percent. So far only three extrasolar planets, all the size of Jupiter, have been discovered in this way.

Experts agree that carbon-based life is most likely to exist on small rocky planets like Earth that orbit a star in orbits similar to Earth's, with a period of about a year. Detecting such a small planet will be a real challenge, however. Just for reference, Jupiter's mass is 318 times that of Earth. The lightest *exoplanet* (extrasolar planet) detected so far has a mass one-tenth that of Jupiter and thirty-three times that of Earth. We have a long way to go.

Planet hunting from satellites is just gearing up as I write this chapter, and all four methods are being applied. The French and European space agencies have joined to build COROT (so named for its scientific goals: Convection Rotations and Planetary Transits). NASA is building the Space Interferometer Mission and is in the early stages of designing the Terrestrial Planet Finder and the Kepler Mission. ESA is also well along in the development of Darwin, a dazzling concept for an interferometer composed of a cluster of satellites.

THE SPACE INTERFEROMETER MISSION

The Space Interferometer Mission (SIM) is essentially a Michelson stellar interferometer, with a 10-meter baseline, that works at visible wavelengths (see chapter 10). Like GAIA, SIM will use the diameter of the Earth's orbit as a baseline from which to measure the parallaxes of stars. Unlike GAIA, which will map billions of stars, SIM will point to a selected set of perhaps thirty thousand. SIM will therefore be able to carry out a broader scientific program.

If NASA is able to overcome a number of technical challenges, SIM would be able to fix the location of a star, relative to a grid of neighboring stars, with an uncertainty as small as 1 micro-arcsecond. That's the angular width of a flea on the Moon. That's also about thirty times smaller than any of the ground-based interferometers might achieve and ten times smaller than GAIA's limit. Such precision would enable SIM to obtain the distances of stars as far away as 75,000 light-years, three-quarters of the diameter of the Milky Way. With long exposures, SIM could reach stars as faint as magnitude 20.

SIM has several scientific goals. Its primary task is to improve the accuracy of the distances to a set of reference stars, especially the hundred nearest Cepheids on which the cosmic distance scale is based. Second, by searching for periodic wobbles in the position of stars, SIM might detect planets. Calculations suggest that SIM should be able to detect any Earth-sized planets around the nearest two hundred stars and any Neptune-sized planets around the nearest two thousand. Third, SIM could enable a detailed study of the internal motions of the galaxy. For example, it could improve the profile of rotation near the galactic center. That might reveal the distribution of dark matter in the largely unexplored region close to the center of the galaxy. Our galaxy, it turns out, is a *barred spiral,* like Messier 83 (fig. 12.6). We still know very little about the motions and distribution of mass in this newest feature, and SIM should be able to help.

SIM's interferometer has modest 30-cm mirrors, but the precision of its optical train is by no means modest. The elements of its interferometer must remain in position to within a nanometer over a five-year flight. (A nanometer is the width of ten hydrogen atoms laid side by side.) Its controllers on the ground must be able to measure the locations of critical elements to this accuracy and to adjust them as needed. These technical requirements pose real challenges to the success of the mission. The Jet Propulsion Laboratory and the Lockheed Martin Corporation share the responsibility for developing the technology.

The SIM science team includes some of the most experienced scientists in the field. Michael Shao, one of the godfathers of the navy's NPOI interferometer on Mount Wilson, chairs the team. He is joined by a dozen researchers with a variety of skills and interests. For example, Geoffrey Marcy (University of California, Berkeley) and Charles Beichman (Caltech) are avid planet hunters. Edward Shaya (University of Maryland) plans to combine SIM's dis-

Fig. 12.6. Our Milky Way galaxy apparently has a "bar" at its center, similar to the one in Messier 83. This galaxy is located in the southern constellation Hydra.

tances and motions to study the formation of the local cluster of galaxies. And Brian Chaboyer (Dartmouth College) is passionate about globular clusters at the boundaries of our galaxy. He hopes to establish their ages and distances more precisely.

NASA plans to launch the SIM in 2009 to an Earth-trailing orbit around the Sun. We wish the team good hunting.

KEPLER RIDES AGAIN

The Kepler Mission is NASA's near-term entry into the race to find rocky, Earth-sized planets in habitable orbits. This satellite will use quite a different principle from either GAIA or SIM. It will search for the slight periodic dimming of a star that occurs when a planet passes across its unresolved disk. From the amount of dimming one could determine the diameter of the planet and from the frequency of the transits, the period of the orbit. Then, estimating the mass of the star from its spectral type, one could determine the radius of the orbit from Kepler's third law, and calculate the temperature of the planet (note 12.5). Finally, ground-based telescopes could search for an atmosphere, and one could decide whether it might support life.

As we mentioned earlier, Kepler would have to lie in the plane of the planet's orbit for a transit to be visible. That is a severe limitation of the photometric method, but with thousands of stars to examine, Kepler might discover many planets this way.

The idea behind this method of finding planets dates back at least to 1971. Frank Rosenblatt, a scientist at Cornell University, proposed to set up a trio of small ground-based telescopes to monitor a large number of stars with CCD detectors. If all three telescopes independently observed a dimming, despite the local differences in seeing, a planet might be the cause. As a check, he suggested observing in two colors, to search for the characteristic reddening at the edge of a star's disk (note 12.6).

In 1984 William Borucki and Audrey Summers (NASA Ames Research Center) reexamined Rosenblatt's idea. They pointed out that a meter-class observatory in space could avoid many of the problems one faces on the ground, but that detection of an Earth-sized planet would still require considerable improvements in the precision of photometers. Over the next decade, large CCD detectors with the required sensitivity and uniformity became available. In the 1990s Borucki revived the idea of a satellite wholly dedicated to discovering planets.

In 1998 Timothy Brown and David Charbonneau (High Altitude Observatory, Boulder, Colorado) initiated the STARE project to detect planetary transits. They first set up a 10-cm telescope and a CCD to monitor the brightness of forty thousand stars simultaneously. Their survey didn't turn up a confirmed transit in three observing seasons, so in 2000 they shifted to the University of Hawaii's 2.24-meter telescope on Mauna Kea. They tested their method on a star (labeled HD209458 in the famous Henry Draper catalog) whose planet had already been found with the radial velocity method. Indeed, they were able to confirm the discovery of a Jupiter-sized planet. And Rosenblatt's prediction of a color change during transit was also confirmed. When they repeated their observations of HD209458 with the Hubble Space Telescope they demonstrated that photometry was now precise enough to detect Earth-sized planets from space.

In January 2001 NASA approved the Kepler Mission, with the goal of finding planets by the transit method. William Borucki is the principal investigator and chairs the high-powered science team. By 2003 the team had defined the parameters of the mission. Kepler will have a 0.95-meter Schmidt telescope equipped with a wide-field photometer. At the back end of the telescope, forty-two CCDs are joined into an array to cover a 10-degree field of view.

Kepler will monitor the brightness of one hundred thousand stars in the constellation Cygnus. An exposure will be made every fifteen minutes over the whole four-year mission. The photometer is state of the art; it is capable of detecting a dip in brightness of 20 parts per million for a magnitude-12 solar-type star during a six-hour transit. Kepler is scheduled for a final critical design review in September 2005. NASA plans to launch Kepler in a heliocentric orbit in October 2007. If Earth-sized planets are as common as some astronomers predict, Kepler should find them by the hundreds.

Meanwhile, more than thirty different groups of ground-based astronomers continue to search for exoplanets. In May 2004 a European team of astronomers discovered two planets with the transit method. The team had actually been searching for something else. They were primarily interested in so-called micro-lensing events. Such an event occurs when a massive object passes between a star and the observer; the object's gravity focuses the light and the star appears brighter. In four years of monitoring some 155,000 stars for these events at the Las Campanas Observatory in Chile, the team discovered 137 cases in which the star *dimmed* slightly.

To rule out the possibility that the dimming is caused by a stellar companion, two French astronomers measured the radial velocity variations of forty-one of the candidate stars, using one of the 8.2-meter telescopes of the VLT.

Et voilà! All but two stars turned out to be binaries. The French team concluded that the remaining two have Jupiter-sized planets. They must be really hot, because they orbit their stars in only a couple of days.

TERRESTRIAL PLANET FINDER

By now you should have some idea of how difficult it is to discover Earth-like planets in orbits favorable to life. NASA is therefore hedging its bets by planning yet another satellite, the Terrestrial Planet Finder (TPF). (No doubt this cumbersome name will be exchanged in the future for the name of some eminent scientist.)

Charles Beichman of Caltech's Jet Propulsion Laboratory chairs the Science Working Team for the TPF. Their basic goal is to capture an *image* of a planet, rather than search for some indirect signature, such as radial velocity or transits. The basic problem with this idea is the enormous difference in brightness between a star and its planet. Depending on the circumstances, the difference could be a factor of a million or more. How could one see the faint planet despite the intense glare of the star?

Two approaches seem attractive. The TPF could carry either a coronagraph for infrared light or a nulling interferometer. A *coronagraph* is a specialized telescope used by solar astronomers to create an artificial total eclipse and to observe the faint solar corona in broad daylight. An opaque element in the telescope's optics blocks out the light of the Sun's disk and allows only the light of the corona to pass to the focal plane. The outer corona is more than a million times fainter than the disk of the Sun, so observing it is just as difficult as finding a planet in the glare of a star. Coronagraph technology is mature and could be applied to a search for planets.

Nulling interferometry is a technique invented back in 1978 by Ronald Bracewell, a famous Australian radio astronomer. To understand its principle, consider a standard Michelson stellar interferometer that produces interference fringes with the light of a distant star (see chapter 10). If an appropriate delay (equal to half a wavelength of light) is inserted in one of the arms of the interferometer, the star's fringes will be "annulled" by destructive interference. Any fringes that remain will arise from a close companion, and with suitable means an image of the companion (hopefully a planet, not a low-mass star) could be recovered. Scientists under contract to NASA are working to reduce

the starlight fringes by factors of tens of millions. The problem becomes easier at infrared wavelengths, where a solar-type star is relatively faint and the
chance of detecting a cool planet is improved.

Designs for the TPF are still at a very early stage. NASA has begun a two-
year study of the two methods and will choose one in 2006. Then the project
may proceed, with a launch possible early in the next decade.

DARWIN

The European Space Agency will not be outdone by anyone in finding and
studying extrasolar planets. As we have seen, the GAIA project is well under
way and stands a good chance of discovering at least Jupiter-sized planets by
means of astrometry. In addition, ESA is developing preliminary designs for
an infrared nulling interferometer called Darwin. Darwin's goal is to find, image, and analyze Earth-sized planets, and to tackle several other astrophysical
problems.

Interferometry with a single satellite is difficult enough, considering all the
factors that can misalign the optics and disturb the pointing. Therefore one
has to admire the audacity of ESA's designers. If they can work out the details,
Darwin would consist of a cluster of six telescope-bearing satellites, flying in
a tight formation to provide as many as fifteen nonredundant baselines, plus
a central hub satellite. An eighth satellite would handle all the communications
to and from the cluster (fig. 12.7).

The six satellites would each carry a classical 1.5-meter Cassegrain telescope.
Their beams would be directed to a central hexagonal hub satellite and allowed
to interfere. Optical delays would be inserted to annul the fringes of a star, in
order to search for planets. Or, in a so-called imaging mode, the six telescopes
might act as a single larger telescope. Few specifics are available as yet, but the
team hopes to resolve details ten times smaller than the James Webb Space
Telescope could—say, 10 milli-arcseconds at 5 microns.

The most challenging problem in such a design is keeping track of the mu-
tual distances of the six telescopes to within a fraction of a wavelength of light.
At ground-based interferometers like NPOI this is done with laser technology
and fringe counting. Presumably the same approach would be applied for Darwin. But in space the difficulty of the problem is amplified a hundredfold. ESA
will have to crack this nut before Darwin can be considered feasible.

Fig. 12.7. An artist's conception of Darwin's fleet of satellites flying in formation

To gain some experience with satellites flying in formation, ESA plans to launch the LISA-Pathfinder in June 2007. It will consist of two free-flying satellites. ESA aims to determine whether the distance between the satellites can be maintained to within 1 nm with precision thrusters. If the experiment is successful it will literally blaze the path for the Laser Interferometer Space Antenna (LISA), an ambitious gravitational wave interferometer (note 12.7). LISA would consist of three satellites, spaced 5 million km apart, to a precision of one-hundredth of a nanometer, or one-tenth the width of a hydrogen atom. Mind-boggling!

If all goes well, Darwin could be launched to the L2 Lagrangian point around 2014. We will look for it.

EPILOGUE

So where do we go from here?

We have come a long way from Galileo's little spyglass. Each increment in the size and power of telescopes has revealed more aspects of the strange and wonderful universe in which we live. We can expect the trend to continue, as long as the nations of the world are willing to cooperate on the vast ventures that lie ahead.

What might we see in the next fifty years? One possibility would be an optical version of the Arecibo Telescope, a transit instrument that points straight up and is designed primarily for spectroscopy. It might have a segmented spherical mirror 300 meters in diameter. Like the Hobby-Eberly Telescope, it would need a complex corrector assembly at the prime focus. It could scan a large part of the sky, at a site strategically chosen to maximize its productivity. How one shields the mirror from sunlight and the weather is beyond me, however.

With sufficient money, I suppose we could build a super infrared interferometric array, with a dozen 8-meter mirrors equipped with adaptive optics. It would be the ultimate cosmic array, capable of micro-arcsecond imaging of distant galaxies, at infrared wavelengths.

Why not an astronomical base on the Moon? This idea has been floating around since at least the Apollo program. With no atmosphere and a gravitational field one-sixth that of Earth's, the Moon is a tempting site. Roger Angel, a familiar name by now, testified before a U.S. Senate committee in November 2003. He proposed to locate an observatory at the lunar South Pole, in the Shackleton Crater, where the Sun never shines and the kelvin temperature is in the low teens. This would be an ideal site for a large far-infrared telescope. It could see only the southern celestial hemisphere, but there is more than enough to see. Serious proposals have also been advanced to assemble such an observatory with robots.

Angel and his colleagues at the Steward Observatory have also proposed to build a 100-meter space telescope. Its parabolic segmented mirror would consist of thin flat membranes. The segments would be kept in alignment by actuators at their edges. Pie in the sky? Not coming from a recognized instrumental genius.

While it's possible that one these schemes may come to pass, it seems more likely that we will see astronomical machines undreamed of now. Stay tuned!

NOTES

1. THE FIRST THREE HUNDRED YEARS

1.1 The Ecliptic and the Meridian

The *ecliptic* is the plane of the Earth's elliptical orbit about the Sun. If one imagines that a large sphere surrounds the solar system, the intersection of this plane and the sphere is a circle on the sky, also called the ecliptic. The Sun appears to move along the ecliptic against the background stars, making a circuit in a year.

The *meridian* is a circle on the sky that passes through the north point on the horizon, through the celestial pole, up through the zenith, and through the south point on the horizon.

1.2 Retrograde Motion of the Planets

Planets such as Mars and Jupiter, with orbits larger than Earth's, revolve around the Sun more slowly than Earth. We have the inside track, so to speak. When Earth catches up and passes them, they appear to reverse their motion against the distant stars, from easterly to westerly. The effect is purely one of parallax, but it makes sense only if you subscribe to a heliocentric view of the planetary system.

1.3 Kepler's Three Laws of Planetary Motion

First, the planets all follow elliptical orbits, with the Sun at one focus of the ellipse. Actually, with a few exceptions, the planetary orbits are very close to circles. Second, the line joining the ellipse's focal point and the planet sweeps out equal areas in equal amounts of time. Thus the planet is revolving faster when it is nearest the Sun (at *perihelion*) than when it is farthest away (at *aphelion*). Finally, the square of the period of revolution is proportional to the cube of the length of the semi-major axis of the ellipse. Thus Jupiter, 776 million km from the Sun on average, revolves in a period of 4,300 days, while Mars, 228 million km from the Sun, completes its revolution in only 690 days.

1.4 The Precession of the Equinoxes

The Earth is like a spinning top whose axis tends to point in a fixed direction toward the distant stars. Because of the gravitational tug of the Sun on the Earth's equatorial bulge, however, the Earth's axis wobbles in a circle. This is called *precession*. A complete revolution of the axis takes about twenty-six thousand years. So in thirteen thousand years the North Star will be some star other than Polaris. Another effect of precession is that the starting dates of spring and autumn change gradually over time. Spring officially starts when the Sun, in its apparent motion against the distant stars, passes from the southern to the northern celestial hemisphere. This event is called the *vernal equinox*. Hipparchus noticed that the exact moment of the start of spring, as marked by the first appearance at sunset of a particular star, came a little earlier each year.

1.5 The Basic Idea of a Telescope

The first telescopes were refractors and used lenses. A lens is in effect a curved prism. A ray of light entering the lens at an angle to its front surface is refracted, or bent away from its original direction and bent back again as it exits the back surface. A convex or plano-convex lens focuses parallel beams of monochromatic light to a common point. It produces a *real* image at the focus, one that can be projected onto a screen. A concave or plano-concave lens, on the other hand, expands parallel beams into a diverging conical bundle. It forms a *virtual* image, visible to the eye or camera but apparently looming out in space in front of the telescope. Figure 1.2 shows how a simple telescope focuses the rays from a target. The long-focal-length lens in front (the objective) forms a real image for the short-focal-length eyepiece to magnify.

1.6 Galileo and Saturn's Rings

When Galileo looked at Saturn with his crude little telescope he saw the planet bracketed by two fuzzy bright objects. He was greatly puzzled by this discovery and decided to keep it to himself for a while. He did record the event, however, writing that in effect the planet was a triple object. Two years later the mysterious objects disappeared, leaving Galileo even more mystified. Christian Huygens, a famous Dutch astronomer, solved the puzzle in 1656. Huygens resolved Saturn's rings with his improved telescope and recognized that they periodically appear to tip edge-on. Galileo must have first seen them flat-on and later edge-on.

1.7 Basic Terms in Optics

In *reflection,* say at a shiny flat surface, a beam of light bounces off at the same angle to the surface as it arrived. In *refraction,* as within water or glass, a beam of light is bent, so that the exit angle with respect to the surface is generally greater than the entrance angle. *Diffraction* is entirely different. A parallel beam of light passing through a pinhole spreads out in all directions past the hole as an effect of the wave nature of light. Newton could explain diffraction only by invoking strange forces acting on the "corpuscles" at the edges of the hole, but Huygens's explanation was closer to our present understanding. He envisioned the progression of a wave through a medium by imagining each point on the wave front as the source of a small spherical wavelet. The envelope of all these wavelets determines the position of the parent wave in the next time interval. In passing through a hole or around an obstacle, the wavelets naturally expand around the edges, producing the spreading effect we see as diffraction.

1.8 How to Grind a Mirror

A fine mirror is a work of art and high precision. Many amateurs today enjoy making one in the time-honored way, by hand. The following process works well with mirrors smaller than about 30 cm. It starts with two identical flat circular blanks of glass. One (the tool) is fixed to a solid bench. The other (the future mirror) rests on top of the tool. A coarse abrasive slurry separates the two. Coarse grinding starts by displacing the top blank by about a radius and moving it back and forth under hand pressure along a chord of the tool. After a dozen strokes both tool blank and mirror blank are rotated in opposite directions by a quarter turn. As the process is repeated, the tool takes on a convex spherical shape and the mirror a concave spherical shape. Finer and finer abrasive grits are used to approach a smooth, spherical surface. The trick is to control the depth of the cut to arrive at the desired focal length and to test progress at each stage. For more details, see http://members.shaw.ca/fvas2/mirror.htm.

1.9 Gustav Kirchhoff

Although Fraunhofer measured the wavelengths of thousands of lines in the solar spectrum, he had no idea what they represented. It remained for Gustav Kirchhoff, a German physicist, to establish their significance and to enable future astronomers to determine the physical properties of stars. Working with the chemist Robert Bunsen, Kirchhoff explored the spectra of a number of light elements, such as sodium, potassium, strontium, calcium, and barium. Using Bunsen's now-famous burner to heat chlorides of such substances, they discovered in 1860 that atoms emit uniquely characteristic patterns of spectral lines. They pointed out the possibility of using such patterns, obtained in the laboratory, to identify chemical elements in the stars and planets and to discover new elements. Their work laid the foundations of astrophysics and led eventually to the atomic theory of Niels Bohr. Later, astronomers learned to use spectroscopy to measure the strength, displacement, shape, and polarization of spectral lines and to determine such stellar properties as chemical composition, temperature, rotation, and magnetic field strength.

1.10 White Dwarfs

According to the theory of stellar evolution, a star produces energy in its high-temperature core by fusing light atoms, like hydrogen and helium, to make heavier atoms, like carbon or oxygen. Each fusion leaves a bit of mass left over in the form of energy, mainly x-rays, according to Einstein's formula $E = mc^2$. Each time the star runs out of a particular type of fuel at its core, the core contracts and the star's surface expands. Eventually the star will run out of all possible fuels and will contract into a white dwarf. This hot, tiny star may be the size of the Earth yet have the mass of the Sun. Its interior is extremely dense. A teaspoon of its matter might weigh a ton on Earth.

1.11 The Sun's Outer Atmosphere

The visible disk of the Sun is surrounded by the *corona,* a million-degree (kelvin) outer atmosphere of ionized hydrogen. It is so faint that it can be seen with the unaided eye only during a total solar eclipse. Between the base of the corona and the *photosphere* (the visible "surface" of the Sun) is another layer, the *chromosphere,* with an intermediate temperature of about 10,000 kelvin. Both layers require special equipment to be seen outside of an eclipse. The *spectroheliograph* enables one to see the chromosphere, and a unique telescope, the *coronagraph,* produces an artificial eclipse on demand.

2. THE AGE OF HALE

2.1 The Coelostat

A *coelostat* is a device for directing sunlight or starlight into an optical system, like a fixed telescope, for further analysis. It consists of two flat mirrors. One is tilted so that its plane is parallel to the Earth's axis of rotation, and is slowly rotated at either the solar or the sidereal rate. The other mirror is stationary but adjustable so as to deflect the light reflected from the first mirror to the final set of optics.

2.2 Statistical Parallax

One can determine the distance of a star by trigonometry if one can observe its displacement against more distant stars from the two ends of a long baseline. For nearby stars, the Earth's orbital diameter (twice the so-called astronomical unit) can serve as a baseline. For more distant stars, Ejnar Hertzsprung used the fact that the Sun moves toward the constellation Hercules at the rate of 2.8 astronomical units per year. That motion produces a baseline that increases with time, and this fact allowed him to get fairly accurate distances of Cepheids.

2.3 Shapley's Mistake

Further research by Walter Baade in the 1940s (see note 4.6) showed that Shapley's variables (now called RR Lyrae variables) are intrinsically fainter than classical variables by a factor of two, although they do have similar light curves. This finding meant that the galaxy was indeed larger than previously estimated (twice as big), though not nearly as much larger as Shapley had calculated (ten times as big).

2.4 Hale's Laws of Sunspot Polarity

Sunspots within a solar hemisphere usually appear in pairs with opposite magnetic polarities. As we view it, the Sun rotates from east to west, so the spot closer to the western edge of the Sun is called the *leader* and the other the *follower.* Hale discovered that, at any moment, the leaders in the northern and southern solar hemispheres have opposite magnetic polarities. Moreover, after a period of about eleven years, the leaders in both hemispheres reverse their polarities.

2.5 The Doppler Effect

A source of light has a characteristic spectrum when it is at rest with respect to the observer. When the source approaches, the observer sees a spectrum shifted toward shorter wavelengths (i.e., toward the blue end) by an amount proportional to the source's speed. Conversely, the spectrum of a receding source appears shifted toward the red end. It is *redshifted.* The effect is similar to the rising and falling pitch of the whistle of a passing train. But sound waves require a medium (air) in which to travel, whereas light can propagate in a pure vacuum. Nevertheless, the same relationship holds between the source's speed and the wavelength shift.

2.6 Optical Aberrations

Like your eyes, a telescope mirror may form imperfect images. If the mirror surface is a sphere, not a paraboloid, different parts of the mirror will focus the light from a star at different points along the mirror axis (we say it has *spherical aberration*). As a result, a camera will record a blurred image. Similarly, if the mirror's shape is distorted, it may magnify an image differently in the horizontal and vertical directions. That is called *astigmatism.* If a mirror has *coma,* images of stars are not points or tiny disks but cometlike trails. Unlike lenses, mirrors are free of at least one annoying flaw, *chromatic aberration.* All colors are focused in the same way.

3. NEW WINDOWS ON THE UNIVERSE

3.1 Pawsey's Sea Interferometer

Interference is the interaction of two waves of the same frequency. When the waves overlap so that their maximums coincide and add together, we say that *constructive interference* has occurred. Similarly, when a maximum and minimum coincide, the waves partially cancel through *destructive interference.*

A radio interferometer combines the signal from two or more antennas to determine the location or width of a source. Having only one antenna, Joe Pawsey placed it on a high cliff (Dover Heights) overlooking the sea. The reflection from the sea surface interfered at the antenna with the part of the wave coming directly from the source. This setup is an analogue, at radio wavelengths, of Lloyd's mirror at optical wavelengths.

3.2 Walter Baade and Rudolph Minkowski

Baade and Minkowski were both astronomers at the Mount Wilson Observatory. Baade discovered that stars belong to two different populations. Stars of Population I are found in open clusters like the Pleiades, within the spiral arms of our galaxy. They are relatively young, contain more of the heavier elements, move relatively slowly, and are associated with gas and dust. Population II are metal-poor stars that are located in globular clusters, in the galactic halo and galactic bulge. They move rapidly and are free of gas and dust. Minkowski worked with Baade on the differences among supernovas. Type Ia supernovas show spectral lines of hydrogen; Types Ib, Ic, and II do not. Later it was found that these latter three types involve the collapse of a massive star, while Type Ia involves the accretion of mass from a star to a compact companion.

3.3 Scattering by Dust Grains

Interstellar dust grains are typically a few microns in size and are composed of carbon or silicon compounds. They absorb light and radiate it as heat, at infrared wavelengths. They also scatter blue light more effectively than red light, causing distant objects to appear reddened. Radio waves longer than a few centimeters are hardly scattered at all, so the galaxy is far more transparent at centimeter wavelengths.

3.4 Origin of the 21-cm Line of Atomic Hydrogen

Hydrogen atoms consist of an electron in orbit about a proton. Each particle has a type of spin. The atom has a slightly higher energy state when the spins are parallel than when the spins are antiparallel. When the electron spin flips spontaneously to the lower energy state, the atom emits the difference of energy as a photon with a wavelength of 21 cm. This is a rare event in space, but great masses of hydrogen can produce a detectable signal.

3.5 Olbers's Paradox: A Clue to the Nature of the Universe

In 1826 the German astronomer Heinrich Olbers asked a simple question: Why is the night sky dark? If the universe is infinite and uniformly filled with stars, every line of sight should end on the surface of a star. Although the light from a star decreases with the square of its distance, the number of stars within a spherical shell also increases with the square of the distance of the shell. The two effects should cancel one another, and we should see a sky as bright as the surface of the Sun. Because the sky is dark at night, one of the following implicit assumptions must be wrong: the universe is static, it is eternal, it is infinite, and it is uniformly filled with stars. We now know that at least the first two assumptions are incorrect.

3.6 Origin of the Big Bang Theory

Georges LeMaitre, a Belgian priest and astrophysicist, recognized in 1927 that if the universe is expanding, it must have had a beginning as a much smaller entity. Gamow adopted this point of view in 1948. He and his student Ralph Alpher were interested in predicting the observed abundances of the elements and proposed that the fusion of nuclei at the extremely high temperatures and density of the early universe determine them. Later, Hoyle and his associates William Fowler, Margaret Burbidge, and Geoffrey Burbidge were able to account for most of the elemental abundances by the nuclear fusion within hot stars. But they had to conclude that the lighter elements deuterium, helium, and lithium do in fact require the extreme conditions of a Big Bang.

3.7 Variability and Size of a Source

In order for a source to vary within a short time, it cannot be any larger than the distance a light beam would travel in that time. Otherwise all parts of the object could not participate in the variation. Imagine, for example, a brush fire starting at one edge of a field. If the fire spreads at a constant speed, we see the field light up in a time equal to the field's size divided by the speed.

3.8 The Farthest Quasars

In 2001 the Sloan Digital Survey, using a 3.5-meter telescope in New Mexico, found two quasars at fractional shifts of $z = 6.0$ and 6.2, which correspond to distances of about 10 billion light-years. (A shift of $z = 6$ means that the shift is six times the laboratory wavelength of the spectral line. A hydrogen line at a "rest" wavelength of 122 nm would appear at $7 \times 122 = 854$ nm $= 0.854$ micron.) Quasars are not the most distant galaxies, however. Recently a galaxy was discovered at $z = 10$.

3.9 Synchrotron Emission

An electron moving in a magnetic field experiences a force perpendicular to its motion that causes it to spiral around the direction of the field. This change of direction counts as an acceleration (a change in the velocity vector) even if the electron's speed is unchanged. Such an accelerated electron is forced to radiate at a frequency that depends on its rate of spiraling, and in a narrow beam tangential to the spiral. This emission is called *synchrotron radiation*.

3.10 The Very Long Baseline Array

The VLBA is a chain of ten 25-meter radio telescopes stretching across the United States from the Virgin Islands to Hawaii. It was built by the National Radio Astronomy Observatory. Each station records its observations of the target on magnetic tape, along with the precise local time. The tapes are shipped to the NRAO, where they are combined into an image.

3.11 Gravitational Waves

Einstein's theory of general relativity predicts that when a massive body moves, the change in its gravitational field propagates outward at the speed of light. The faster the motion and the more massive the body, the stronger the wave. Such waves are predicted to be exceedingly weak under the best of circumstances. Supernovas and mergers of black holes are two possible kinds of detectable sources. The Hulse-Taylor result is an indirect proof that they exist. Several groups, at Caltech and MIT, for example, are trying to detect such waves by observing the infinitesimal deformations they produce as they pass through a test object.

3.12 Interstellar Molecules

The present number of known interstellar molecules is about 120. Among them is glycine (CH_2NH_2COOH), an amino acid that is one of the constituents of DNA. That discovery raises the interesting possibility that such interstellar molecules played a role in the beginning of life on Earth—and perhaps elsewhere. Astronomers are searching for signatures of life on planets around other stars.

3.13 Detectors for X-rays and Gamma Rays

Most cosmic x-ray sources are so weak that photons arrive one by one, not in a shower. Detectors are designed to respond to single photons. A *proportional counter* is a cell of gas with an array of internal electrodes. An x-ray photon that enters the cell ionizes a gas atom and releases a free electron, which is attracted to a positively charged electrode. The electron collides with more atoms, which produce more free electrons, and so an avalanche is produced that is more easily detected. The size of the avalanche is a measure of the energy of the incident x-ray photon.

A variety of materials, including plastics, emit visible light when illuminated by an x-ray photon. Such *scintillation* detectors employ a sensitive photomultiplier to convert a single visible photon into a cascade of photons that can then be easily measured.

Scintillators are also used to detect gamma rays. Another type relies on the *Compton scattering effect,* in which a high-energy photon collides with an electron in some medium and gives the electron enough energy to reach an electrode.

3.14 The Launch of Uhuru

Riccardo Giaconni's satellite was launched from San Marco, off the coast of Kenya, on December 12, 1970, the seventh anniversary of the independence of Kenya. Uhuru means "freedom" in Swahili.

3.15 Color and Temperature

As a general rule, the hotter an object is, the shorter are the wavelengths it emits. Conversely, the colder it is, the longer are the wavelengths it emits. So, for example, cold interstellar clouds emit in the far-infrared wavelengths (millimeter and centimeter ranges), while hot young stars emit in the visible and ultraviolet wavelengths. You can see this effect when the electric coils in your toaster heat up. They turn red at first, then bright yellow.

3.16 Detectors for the Infrared

A *lead sulphide cell* changes its electrical resistance when light with wavelengths between 1 and 4 microns illuminates it. The resistance change varies with the strength of the light. These cells are cooled in liquid nitrogen to reduce thermal noise in the cell.

Germanium bolometers are sensitive to all the infrared wavelengths and are hundreds of times more sensitive than lead sulphide. The bolometer is cooled to 4 kelvin in liquid helium. IR light changes the conductivity of the bolometer in proportion to the strength of the light.

A new generation of detectors arrived in the 1990s. *Indium antimonide* and such exotic blends as *mercury-cadmium-tellurium* and *indium-gallium-arsenic* are highly sensitive to IR, out to 5 microns. *Two-dimensional arrays* of IR detectors have recently revolutionized IR research, because they allow imaging of extended sources.

4. THE RISE OF THE GREAT CENTERS

4.1 U.S. National Observatories

The National Science Foundation funds several national observatories: the National Radio Astronomy Observatory, the National Solar Observatory, the National Optical Astronomy Observatory (which include Cerro Tololo Inter-American Observatory and Kitt Peak National Observatory), and the National Astronomy and Ionosphere Center at Arecibo.

4.2 Ritchey-Chrétien Optics

George Ritchey was G. E. Hale's favorite optician. While grinding the 60-inch mirror he and French optical designer Henri Chrétien invented a telescope design with a hyperbolic primary and secondary. This arrangement ensures perfectly round star images even at the edges of the field of view. That is, it eliminates *coma,* the aberration that produces cometlike star images. Hale and Ritchey quarreled over using the new design for the 100-inch, and Hale fired Ritchey. Both the 100-inch and the 200-inch have standard Cassegrain optics, with parabolic main mirrors. All large telescopes since the 200-inch use RC optics, however. Incidentally, Professor Chrétien was the inventor of the special optics used for Cinemascope in the 1950s.

4.3 Signs of Dark Matter

Vera Rubin measured the Doppler velocities of stars at increasing distances from the center of a typical rotating galaxy. She discovered that the stars were moving too fast to be held in their orbits solely by the gravity of the visible mass of the galaxy. That result pointed to the presence of some kind of invisible, or *dark,* mass between the stars. Later work has shown that more than 90 percent of the matter in the universe may be dark.

4.4 The Classification of Stellar Spectra

Edward Pickering, the director of the Harvard College Observatory, initiated a vast project to obtain and classify the spectra of thousands of stars. He hired several women to
draft a classification scheme. Annie Jump Cannon was the third assistant to devise a
scheme, and hers has lasted to this day. It is based on the relative strengths of spectral
lines that form at different stellar temperatures. Her scheme is essentially a temperature
scale for stars. She classified over four hundred thousand stars, which formed the famous
Henry Draper catalog.

4.5 Double-Beam Telescope

To estimate the quality of the seeing at a site, observers usually began by looking at star
images with small telescopes, 10 or 20 cm in size. But a meter-class telescope would reveal poorer seeing under the same conditions, for reasons that we will explain in a later
chapter. In order to get a more realistic estimate of the seeing, the double-beam telescope
was often used. It consists of a 10-cm lens at each end of a 2-meter aluminum beam that
is on an equatorial mount. Starlight from the lenses is directed to the center of the beam
and to a camera. Two long-exposure star trails can be photographed on a single photographic film. The wiggles and blurring of the trails yield quantitative information on the
variation of seeing over a 2-meter mirror.

4.6 Walter Baade

Baade was a Dutch-born American astronomer. While on the staff of the Mount Wilson
Observatory he discovered two different populations of stars in our galaxy. Young stars
were associated with dust in the spiral arms, old stars were free of dust and were found
in a halo around the galaxy. He also discovered that two types of Cepheid variable stars
existed, with different absolute brightness. Previous observers had been unaware of the
difference and used both types as "standard candles" to measure the distances to galaxies. Once Baade's correction was recognized, the estimated size of the visible universe had
to be doubled. See also note 3.2.

4.7 Inversion Layer

During daylight hours the air temperature usually decreases with increasing altitude up
to a height of perhaps 2,000 meters and then begins to rise. This height is called the *temperature inversion layer*. Clouds are normally confined below this height because convection of hot air requires a negative temperature gradient with height. Hence an observatory above the inversion is more apt to be free of clouds.

4.8 Testing Seeing with Star Trails

In the Northern Hemisphere the night sky appears to rotate about the North Celestial
Pole, near the star Polaris. In a long-exposure photograph the stars trace concentric cir-

cles around the pole. Fluctuations in the brightness and width of these star trails are measures of the quality of the seeing, at least for the size of telescope used in testing. A similar technique can be used in the Southern Hemisphere.

4.9 Nasmyth Ports

In an alt-azimuth telescope, the Nasmyth ports are two locations on either side of the azimuth axis where a large stationary instrument can be mounted. The light beam is directed through the altitude axis to the instrument with two flat mirrors. James Nasmyth, a Scottish engineer, invented this convenient arrangement in the nineteenth century.

4.10 The European Southern Observatory's S-Camera

S-CAM is a 6 × 6 array of tantalum-aluminum *superconducting tunnel junctions* (or *Josephson junctions;* see note 5.9) working at a temperature as low as 1 degree kelvin above absolute zero. Each Josephson junction consists of a pair of superconducting films separated by an insulating layer. Each single-incident visible photon releases as many as a thousand electrons within 5 microseconds. In comparison, a charge-coupled device collects only one electron per photon. The number of electrons can be counted and is proportional to the energy of the photon. The output of the camera is sent to a computer that assembles the energy (i.e., wavelength) information into a spectrum.

5. THE HUBBLE SPACE TELESCOPE

5.1 High Spatial Resolution

The sky is not absolutely dark between the bright stars. There is always a faint background of distant galaxies, as well as the light scattered from nearby stars by interplanetary dust (the *zodiacal light*). In order to distinguish a faint galaxy against the background, one wants to gather as much light from it, and *only* it, as possible. That translates into high spatial resolution, or razor-sharp images. All ground-based telescopes are limited in resolution, to some degree, by turbulence in the Earth's atmosphere, while the HST is limited only by the quality of its optics.

5.2 The Diffraction Limit of a Telescope

The angular resolution of a perfectly constructed mirror is determined by the wave nature of light. Light from a distant star arrives in a series of parallel wave fronts, analogous to a train of incoming ocean waves. When a wave front encounters the finite diameter of the mirror (D), it is diffracted as well as reflected. *Diffraction* means that the wave tends to spread out into a small angle of w/D radians, where w is the wavelength of the light (see figure 10.1 for an example).

The mirror brings all the light to a focus, however. As a result of diffraction, a point on the focal plane of the mirror receives light from slightly different directions and hence with different phases. (The *phase* is the fraction of the cycle of vibration. Recall the phases of the Moon.) Waves with different phases may either cancel or add, in a process called *interference*. The result is a circular pattern of light and dark rings (the *Airey disk*) instead of a perfectly sharp point. The diameter of the central bright spot is the *diffraction limit*, the sharpest image of a star this perfect mirror can make. The larger the diameter of the mirror, the sharper the image. Moreover, the shorter the wavelength of the light, the sharper the image.

5.3 Dirty Space

All sorts of things are flying around in space at high speed that could damage the HST. There are meteoroids the size of dust particles and as large as pinheads. There is also *space debris*—nuts and bolts and chips of paint left over from previous space missions. Fortunately the HST has a narrow opening at the front, which greatly reduces the chance of a direct hit on the primary mirror. In addition, this opening can be covered with a hatch when wanted.

5.4 Stellar Magnitudes

The magnitude scale of brightness is logarithmic. A star of magnitude zero is 2.5 times as bright as one of magnitude 1, which is 2.5 times as bright as one of magnitude 2, and so on. The human eye can see down to magnitude 6, and the HST's Wide Field Camera can see down to magnitude 27.5 in visible light, or a factor of four hundred million times fainter. The Near Infrared and Multi-Object Spectrometer (NICMOS), discussed later in chapter 5, does even better, magnitude 30, in infrared light.

5.5 Charge-Coupled Devices

These solid-state detectors of light have largely replaced photographic film for scientific purposes. They consist of arrays of thousands or millions of photodiodes packed onto a chip about the size of a postage stamp. Each diode (or *pixel*) is about 10 microns in width. When light hits a diode, it releases an electrical charge in the semiconductor of the diode. The brighter the light, the larger the charge. The chip contains a circuit that periodically sweeps each diode's charges down to one corner. Diode by diode, the electronic image arrives at an *analog-to-digital converter* that counts the number of electrons in each pixel. In this way the original picture is converted into an array of binary numbers. A great deal of effort has been invested in increasing the number of pixels in a CCD, to provide higher resolution or a bigger picture, and improving the sensitivity of the diodes to ultraviolet and infrared light. CCDs are now common in video cameras and many other consumer products.

5.6 Geosynchronous Orbits

A satellite in *geosynchronous orbit* circles the Earth in twenty-four hours, remaining above a fixed spot on Earth. That is a convenient arrangement for communicating with the satellite. To orbit in twenty-four hours, the satellite has to be at a height of 35,800 km above the Earth's equator.

5.7 A Satellite's Lifetime in Orbit

Friction with the Earth's atmosphere is the principal limiting factor in the life of a satellite. Obviously, the higher the orbit of the satellite, the thinner the atmosphere and hence the longer the projected lifetime of the satellite. Other effects, such as tidal, electric, and magnetic forces, can affect the lifetime also.

5.8 Spherical Aberration

A perfect parabolic mirror focuses all the light from a distant star to a single point at the focus. In contrast, different parts of a perfect spherical mirror focus starlight to different points along the axis of the mirror. These multiple images of the star are defocused when they are projected to the focal plane. The result is a fuzzy image. The slight differences in shape between a paraboloid and a sphere are called *spherical aberration.*

5.9 Josephson Junction Detectors

Brian David Josephson, a British physicist, discovered an unusual effect while studying electrical conductivity at very low temperatures. His junction consists of two superconducting layers separated by a very thin nonsuperconducting layer. At temperatures near absolute zero, electrons can tunnel through the insulating layer without an applied voltage. If a small voltage is applied, the current oscillates at gigahertz frequencies. The amount of current depends very sensitively on an applied magnetic field. These effects can be used to measure very small voltages, magnetic fields, or photon showers.

6. A MIRROR IN MANY PIECES

6.1 Joe Wampler's Design

I haven't been able to learn just how Wampler intended to produce a monolithic 10-meter mirror. Corning Glass Works, as we shall see in a later chapter, gained experience in "slumping" a heated glass disk to fabricate a large, thin mirror blank, but it's not clear that this technique was available in the early 1970s.

6.2 Luis Alvarez and the Great Extinction

Alvarez received the Nobel Prize for his outstanding experimental work in elementary particle physics, but he was interested in a wide range of topics. He developed a theory that a giant asteroid collided with the Earth sixty-five million years ago and caused the extinction of half the life on the planet. He knew that iridium, a silvery metal, is extremely rare on Earth but quite abundant elsewhere in the solar system. An asteroid would carry a large load of iridium. So he looked for a thin layer of iridium-rich clay in the vicinity of the putative impact zone. He found one, dating back to sixty-five million years ago, evidence that supported his conjecture. His theory is still not accepted by all the experts, however.

6.3 Nasmyth Foci

In a Cassegrain telescope with an alt-azimuth mount, a secondary mirror directs the light beam through a hole in the large primary mirror and to an instrument at the Cassegrain focus. If a tertiary mirror is inserted near the Cassegrain focus, the beam can be directed through the horizontal axis to either of two so-called Nasmyth foci, at the side of the telescope. In figure 6.6, the two "ears" that appear at the sides of the left telescope are the platforms for the Nasmyth foci. These platforms rotate in azimuth with the telescope mirror. The focal ratio at a Nasmyth port is typically f/20 or larger. Large, heavy instruments can be mounted semipermanently at these foci.

6.4 Paraboloids and Hyperboloids

Parabolas, hyperbolas, and ellipses are all *conic sections,* curves that one gets by slicing a cone in different directions. A slice across the tip of the cone produces an *ellipse* (a circle is a special case), a slice parallel to the axis of the cone produces a *hyperbola,* and a slice parallel to the side of the cone produces a *parabola.* If one rotates a parabola about its axis one gets a *paraboloid,* a surface in two dimensions (similarly with a hyperbola). Mirrors are often called parabolas, but strictly speaking they're paraboloids. A single paraboloid focuses parallel rays to a point, but it takes a concave and a convex hyperboloid in tandem to do that, as in the Ritchey-Chrétien design.

Nearly all large telescopes built in the past fifty years have this optical design, because it offers a wide field of view that is completely free of coma aberrations (see note 2.6).

6.5 Echelle Spectrograph

An *echelle grating* spreads the light from a source into a series of overlapping spectra. A *cross-dispersion prism* or *grating* can be used to unscramble the overlap, by shifting each spectrum parallel to itself. This combination of echelle and prism forms a stacked array of high-resolution spectra in a compact arrangement that is suitable for detection with a CCD.

6.6 Supernovas

Supernovas are relatively rare. One would appear in a typical galaxy only once or twice in a millennium. The supernova searchers find these elusive beacons by first capturing as many as fifty thousand galaxies in one exposure. Then they make a second exposure in the same direction, say, three weeks later and subtract the one image from the other. In this *difference image* only the supernovas have changed in brightness, and they are easily picked out. Then these candidates are measured repeatedly in several colors to obtain their light curves, in several colors. Type Ia supernovas can be identified by the distinctive slow rise and fall of their brightness.

7. SPINNING GLASS

7.1 Optical Fibers

A beam of light passing through a long, thin fiber is reflected at a grazing angle many times at the fiber walls. Snell's law of refraction ensures that each reflection at an angle less than a certain critical angle is *total*, meaning that no light can escape through the wall to the outside.

7.2 Large Liquid-Mirror Telescopes

Liquid-mirror telescopes have become popular again, even though they can observe only near the zenith (the point directly overhead). NASA built a 3-meter telescope in Cloudcroft, New Mexico, to observe manmade debris in space. It employs 15 liters of liquid mercury in a spinning parabolic dish. A consortium led by the University of British Columbia is building a 6-meter mercury telescope, primarily for spectroscopy of distant galaxies.

7.3 Mirror Seeing

Any source of heat in a telescope enclosure can set up turbulent air drafts that can ruin the seeing. A mirror heats up during the day and cools down by night as the air temperature changes. That too can ruin the seeing, unless the mirror is cooled rapidly to the ambient nighttime temperature before observing starts. A hollow core allows for rapid cooling, typically less than an hour.

8. A PROFUSION OF TELESCOPES

8.1 Zerodur

Zerodur, a hard, brittle material, has a glassy (amorphous) state and a crystalline state. In the crystalline state the material contracts when heated, while in the glassy state it expands. By controlling the temperature and rate of cooling, a mixture of states can be created that neither expands nor contracts with a change of temperature.

8.2 Coudé Focus

A Coudé optical system consists of a concave secondary mirror (often a hyperboloid) and one or more flat diagonal mirrors. They form an image at a fixed position outside the framework of the telescope, where heavy instruments may be mounted permanently.

9. BEATING THE SEEING

9.1 Speckles and Interference

Another way to think about the formation of a speckle is to consider each wrinkle in an incoming wave front as an independent wave front (a "wavelet," if you will) that moves in a slightly tilted direction. All the wavelets moving in the same direction arrive at the same position on the telescope's focal plane; those that move in another direction arrive at slightly different positions. These different wavelets then interfere to produce the cloud of speckles we see. Hence the term *speckle interferometry.*

9.2 Stellar Interferometer

Albert Michelson mounted two pairs of diagonal mirrors on a long beam at the front end of the 100-inch telescope's tube. The mirrors relayed the light from a star to the main mirror, and the two (monochromatic) beams were allowed to interfere at the focal plane. A pointlike star produces a sinusoidal interference pattern of highly contrasting bright and dark bands. Any broad object, like Betelgeuse or a planet, produces bands of lower contrast, because of the destructive interference of waves from different parts of the object. The "visibility" of the bands is a measure of the width of the source and can be calibrated to yield quantitative sizes.

9.3 Continuous-Faceplate Mirrors

Hundreds of actuators can now be formed on a single slab of piezoelectric ceramic. Each actuator expands by the required distance when the appropriate high voltage is applied to its electrode. This type of mirror is expensive to fabricate, however, and is employed only on the largest astronomical telescopes, where the effects of hundred or thousands of air cells have to be removed. The 8-meter Gemini North and 10-meter Keck telescopes use such deformable mirrors, for example.

9.4 The Latest in Solar Adaptive Optics

In June 2003 Thomas Rimmele and his team at the National Solar Observatory demonstrated an AO system that corrects seventy-six segments of a wave front in real time. It is based on the Shack-Hartmann sensor and corrects a portion of the Sun's image twenty-five hundred times per second. One segment of the image is chosen as a reference, and all other segments are corrected to match it. The system reached the diffraction limit (0.1 arcsecond) of the 0.76-meter Dunn Solar Telescope at a wavelength of 450 nm. This system is a prototype for a proposed 4-meter Advanced Technology Solar Telescope.

9.5 Reconstructors

A *reconstructor* is basically a fast, specialized computer that calculates the signals to the deformable mirror. Different mathematical algorithms are used for this purpose, depending on the type of mirror. In a continuous-faceplate mirror, like the one used in Come-On, the measured wave-front slopes are canceled by arranging the heights of all the mirror's actuators. Each actuator affects not only the area directly around it but also, to a small extent, the areas on all sides of it. Because the actuators interact in this way, the reconstructor must solve for *all* the actuator voltages simultaneously. In practice that means solving as many simultaneous linear equations as there are actuators. Each equation expresses a single measured wave-front slope as a combination of all the actuator heights or, equivalently, actuator voltages. Matrix multiplication is the mathematical tool of choice for solving this system of equations, but many refinements have been introduced to overcome the unavoidable errors and gaps in measurements.

9.6 Roddier's Prototype Curvature System

This analog device uses only electrical currents to process the image and performs no real-time digital calculations. It consists of three main components, a flux detector, a vibrating membrane mirror, and a bimorph mirror. The *flux detector* has an array of thirteen small lenses that form images of different parts of an incoming wave front on a corresponding array of photodiodes, as in a Shack-Hartmann sensor. To switch between the two out-of-focus planes, the *membrane mirror* cycles rapidly between two states, concave and convex. When it is concave, the plane in front of the true focal plane is directed to the sensor, and when it is convex, the other out-of-focus plane is sampled. The diode currents in the two states are subtracted in real time and become the inputs to the bimorph mirror. In the prototype, the *bimorph* was divided into thirteen radial segments to match an equal number of segments on the wave-front sensor and on the wave front. Each segment has an electrode that connects the mirror to a corresponding element of the sensor. White light was used in the sensor, and the science camera recorded infrared light.

9.7 Stellar Magnitudes

The visual brightness of astronomical objects is measured on a logarithmic scale called a *magnitude scale*. If two stars differ in brightness by a factor of 2.5, their magnitudes differ by 1. If they differ in brightness by a factor of 100, their magnitudes differ by 5. Fainter stars have larger magnitudes, contrary to intuition. Thus Vega has magnitude 0.14 while fainter Regulus has magnitude 1.34.

10. THE ASTRONOMER'S MICROSCOPE

10.1 Ryle's Beam

As you can see from figure 10.3, the width of Ryle's beam varied with the angle above the horizon or, equivalently, in the south-to-north direction on the sky. That meant that the fringe separation was smaller closer to the horizon. So a source that passed through the beam at low altitudes produced a sinusoidal tracing with a shorter wavelength than a source at higher altitudes.

10.2 Scanning a Wide Strip of the Sky

At each position of elements A and B, their beams can be pointed in different directions in the sky to scan for sources. Pointing is done by introducing an electronic time delay in one arm of the interferometer, as described in figure 10.4.

10.3 The Spatial Frequency Plane

As seen from the source, the length and orientation of a projected baseline corresponds to a spatial wavelength in two orthogonal directions, say, x and y. The reciprocals of these wavelengths (1/wavelength) are the spatial frequencies u and v. If we plot u and v as a source rotates across the sky, we see that they trace an ellipse in the u,v plane (fig. 10.7). Now if our interferometer has several antennas, with nonredundant baselines, each baseline traces its own ellipse. In this way a large range of u,v pairs are sampled in an area whose maximum size corresponds to the longest baseline. A parabolic dish with a diameter equal to this maximum baseline would sample all the spatial frequencies and would yield a high-quality image of the source. The array does a fairly good job by sampling a "sufficient" number of u,v pairs.

10.4 An Independent Discovery of Aperture Synthesis

Australian physicists William Christiansen and John Warburton were also using Fourier methods, but with a twist. They wanted to obtain two-dimensional images of the Sun at a wavelength of 21 cm. So in 1953 they built two independent interferometers situated at right angles to each other. Each one consisted of a row of many fixed antennas that produced a row of narrow, fan-shaped beams. One interferometer's beams were narrow in the east-west direction, and the other set were narrow in the north-south direction. As the Sun drifted through the multiple beams for a few hours around noon, the observers obtained one-dimensional scans in many directions on the solar disk.

Each scan contained information on a range of spatial wavelengths, and by combining scans from many days using Fourier methods, they obtained an image of the Sun. At a radio wavelength of 21 cm this image had a resolution of about one-tenth of the solar disk, or 3 arcminutes—very good for its time. Unlike O'Brien, these observers had no need to change the spacing or the orientation of their antennas. In effect the rotation of the Earth changed it for them. They seemed to be unaware that they had demonstrated a new principle, aperture synthesis, which Ryle and Hewish described with greater clarity.

10.5 Phase Closure

The Earth's ionosphere is a set of layers of partially ionized atoms at altitudes between 80 and 250 km. These layers act to retard and scatter cosmic radio waves. They introduce phase errors, f_1 and f_2, in each arm of a two-antenna interferometer. So the phase difference D_{12} between the antennas contains the difference of the true phases $t_{12} = t_1 - t_2$ plus a random error $(f_1 - f_2)$ or $D_{12} = t_{12} + (f_1 - f_2)$.

Suppose one has three antennas from which one can form three two-antenna arrays. Then if one adds the phase differences one has $D = D_{12} + D_{23} + D_{31} = t_{12} + t_{23} + t_{31}$, in which the random errors have all canceled. This "closure phase" D can be used to reconstruct a source distribution with higher reliability.

10.6 Very Large Telescope Interferometer Refinements

When a pair of 8-meters are coupled as an interferometer, the fringes they produce can still wander enough to spoil a measurement. That is particularly serious with faint targets, which require long exposure times. Therefore the ESO team is building a dual-beam facility, called Phase Reference Imaging and Micro-arcsecond Astrometry (PRIMA), that uses a bright guide star to stabilize the fringes of a nearby faint object. The idea is similar to that used in adaptive optics. When complete, PRIMA should enable the VLTI to image objects as faint as visual magnitude 19, which would include many distant galaxies.

11. TOWARD THE EVER-RECEDING HORIZON

11.1 Gravitational Lensing

Lensing is an effect caused by the distortion of space-time in the vicinity of a massive object, as predicted by Einstein's theory of general relativity. In *strong* lensing, light from a distant galaxy that grazes a foreground galaxy is focused into multiple images, or arcs, or a ring. These subimages are markedly enhanced in brightness. The mass of the foreground galaxy can be determined from the pattern of the images.

Weak lensing occurs among a large cluster of galaxies. It is caused by mass inhomogeneities that induce a *tidal* distortion of light rays, in which the deflections vary with angle on the sky. This effect is much weaker than the uniform deflections one sees in strong lensing. The light rays from neighboring galaxies undergo a similar tidal distortion. Because the distribution of galaxies within a large volume should be random, any departure from randomness yields information on the tidal field of force and, in turn, on the distribution of mass.

11.2 The Search for Extra-Terrestrial Intelligence

Frank Drake was the director of the 305-meter Arecibo Radio Telescope in the 1970s. Using his famous eponymous equation, he estimated a nonzero probability for the existence of intelligent civilizations in other galaxies. As director, he initiated a Search for Extra-Terrestrial Intelligence (SETI), listening for nonrandom radio signals from space. During the past forty years he has pioneered efforts to detect such intelligent life by increasing the number of radio frequencies one scans continuously. His latest venture is the Allen Radio Array, which will be a dedicated array of 350 6-meter radio dishes for SETI as well as innovative astronomy.

11.3 Origins

In 1993, two years after the launch of the HST, NASA asked the Associated Universities for Research in Astronomy to convene an advisory committee to consider what to build next. A group of eighteen distinguished astronomers, from as many institutions, reviewed recent progress and opportunities for research and presented a vision of a grand theme of exploration. They called it Origins. It would be a broad program, aimed at understanding the origin and evolution of the universe. As part of this ambitious program, they proposed to search for distant planetary systems that might support life as we know it, a subject that would have great appeal for the general public.

Their vision encouraged NASA to propose to build the Next Generation Space Telescope, the successor to the HST. Initially they suggested a 4-meter telescope. When the Hubble Deep Field of 1995 showed that the earliest phases of galaxy formation might be glimpsed with a larger telescope, they raised the target to an 8-meter. With budget cutbacks, the NGST gradually evolved into the 6.5-meter James Webb Space Telescope (JWST), named for the NASA administrator who presided over the Apollo program.

11.4 Lagrangian Points

L1 and L2 are two positions along a radius from the Sun to the Earth. L1 lies on the day side of the Earth, at about four times the distance of the Moon; L2 lies at about the same distance on the night side. At either location the gravitational forces of Earth and Sun combine in such a way that a satellite would orbit the Sun in the same time as the Earth does. At the L2 point the JWST would keep pace with the Earth and would remain permanently in its shadow.

11.5 The Multi-Aperture Scintillation Sensor

MASS consists of a small off-axis reflecting telescope and a detector at the prime focus. Starlight passes through a mask with four concentric circular rings in front of the telescope. A detector records the amount of twinkling of the light through each ring. The difference in the amount of twinkling between two rings can also be recorded. In fact, the rings can be paired in six unique ways and their differences recorded. From these data the contributions of six different layers to the total scintillation can be computed in real time by a computer.

12. THE FUTURE IN SPACE

12.1 Membrane Mirrors

Roger Angel and his colleagues at the University of Arizona proposed in 2000 to build enormous ultra-lightweight mirrors from a stretched reflective membrane. The mirror would presumably unfold from its spacecraft like an umbrella. In an improved version, the membrane could be shaped into a long-focus paraboloid by electrostatic forces between the membrane and an array of control electrodes. I don't know whether anyone is considering building one, but it's a clever scheme.

12.2 Heterodyne Spectroscopy

With suitable electronics, radiation at millimeter, submillimeter, and far-infrared wavelengths could be amplified and analyzed with the same methods one uses for radio wavelengths. One obstacle to this approach is the lack of low-noise amplifiers at frequencies higher than 300 GHz. To avoid this problem, the cosmic signal is mixed *(heterodyned)* with a lower frequency from a tunable laser. The result is a signal at an intermediate frequency that can be amplified and analyzed. Because the laser is tuned sharply in frequency, the IF signal is narrow-band. In effect the process selects a narrow band of frequencies for further analysis, just as a spectrograph does.

12.3 Hipparchus the Greek

The acronym Hipparcos was chosen to honor Hipparchus, one of the giants of classical Greek science. With little more than keen eyesight and intelligence, he discovered the precession of the equinoxes. See note 1.4.

12.4 Parallax

The *parallax* of a cosmic object is the angular change in its position on the sky when it is observed from opposite sides of the Earth's orbit. The diameter of the orbit (150 million km) can then be used as the base of a long thin triangle whose base angles are known from the measured parallax. The distance to the object follows directly from elementary trigonometry. Parallaxes are small angles. Proxima Centauri, the nearest star, has a parallax of only 0.76 arcsecond. Before space astrometry was possible, only distances closer than about 1,000 light-years could be determined with this method, because the uncertainty in the parallax was comparable to the parallax itself.

12.5 Kepler's Third Law and the Diameter of Planetary Orbits

Kepler's third law states that the square of a planet's period is proportional to the cube of its orbit's semi-major axis. Sir Isaac Newton calculated the constant of proportionality from basic physics. So if one determines the period of a planet from the frequency of the stellar eclipses it produces, one can find the orbit's size, in angular measure. Then, if one also can measure the radial velocity, or if one knows the distance of the star, one can determine the orbit's size in kilometers. Knowing how close the planet gets to the star, and the star's surface temperature, one can estimate the planet's surface temperature and decide whether life is likely to exist there.

12.6 Limb Darkening of Stellar Disks

The edges *(limbs)* of a star's disk are generally cooler and therefore redder than the center of the disk. As a dark planet covers a portion of the limb, at the beginning and end of a transit, the contribution of the limb to the red part of the spectrum decreases. So the planet produces a subtle change in the spectrum or color of the star. That is a useful signature of a planetary eclipse. If the star had a stellar companion, not a planet, one would see quite a different change in the spectrum and brightness of the system.

12.7 Gravitational Waves

When a massive object shifts position rapidly, its gravitational field must change throughout space accordingly. Einstein predicted that the change in the field is propagated from the object in the form of a gravitational wave. Joseph Taylor and Russell Hulse discovered a pulsar as a member of a binary system. The pulsar's orbital period was initially 7.75 hours but decreased continuously over the following decade. The two scientists showed that the system was losing energy by the radiation of gravitational waves, in full accord with the theory of general relativity. They won the 1993 Nobel Prize in physics for their work.

Now astronomers are trying to detect gravitational waves caused by such catastrophes as supernovas and the final collapse of neutron binary star systems. A dozen independent groups are working with ground-based detectors, and several space-based systems, such as NASA's Laser Interferometer Space Antenna, are being developed as well.

GLOSSARY

accretion disk
: A thin rotating disk of gas that surrounds a gravitating body, such as a star or black hole. The gas originates in a neighboring body, such as a star or cloud, and settles at a distance from its host where gravitational and centrifugal forces are equal.

achromat
: An optical element (lens or mirror) that affects all colors (wavelengths) the same way.

achromatic
: Independent of color.

altitude
: The angle of an astronomical object above the horizon.

altitude-azimuth mount
: A telescope mount with an axis for rotation vertically (in altitude) and an axis for rotation horizontally (in azimuth).

angular size
: The width (in degrees, arcminutes, or arcseconds) of a distant object.

arcminute
: An angle equal to 1/60 of a degree of arc.

arcsecond
: An angle equal to 1/60 of an arcminute, or 1/3600 of a degree of arc.

azimuth
: The angle, measured horizontally, from the direction of due south.

Cassegrain
: A telescope with a parabolic primary mirror and hyperbolic convex secondary. The prime focus lies behind the primary.

charge-coupled device (CCD)
: An array of solid-state photosensitive units. In sizes as large as $2,000 \times 2,000$ pixels, they have largely replaced photographic cameras in astronomy.

chromatic aberration
: A defect in an imaging system that causes light of different wavelengths to focus at different positions.

coelostat
: An optical device used to direct light from an astronomical object to a fixed telescope. It consists of two flat mirrors, one that rotates and another that is stationary but adjustable.

coma
: An optical aberration in which star images are elongated, like comets.

Coudé focus	An auxiliary focal point of a telescope, produced with a convex secondary mirror and one or more flat diagonal mirrors.
declination	An angle on the celestial sphere, similar to latitude on the Earth.
degree	An angle equal to 1/360 of a circle.
diffraction	The spreading of a wave as it passes around an obstacle or through a hole.
diffraction limit	The limiting angular resolution of a telescope, which is determined by the ratio of the wavelength of the light to the diameter of the mirror.
Doppler shift	The wavelength shift of the spectrum of a source that is moving toward or away from the observer.
equatorial mount	A telescope mount in which one of the two axes is parallel to the Earth's axis.
figure	To grind a mirror to the required shape.
focal ratio	The focal length of a lens or mirror divided by its diameter. Smaller ratios lead to shorter exposures of a given source of light.
frequency	The number of times per second (or other unit of time) an event recurs. Or, the reciprocal of the period.
frequency, spatial	The number of times per meter (or kilometer) a certain wavelength recurs. Or, the reciprocal of the wavelength.
Gregorian	A telescope with a parabolic primary and an elliptical concave secondary that lies in front of the prime focus.
hyperbola	A curve obtained by slicing a cone parallel to its axis.
hyperboloid	A surface created by rotating a hyperbola about its axis.
interference	The interaction of waves when superposed. When maxima coincide, they add constructively; otherwise they interfere destructively.
interferometer	A device that combines two or more waves and allows them to interfere.
kelvin	A unit of temperature equal to the Celsius degree. The zero-point of the kelvin scale is absolute zero, or $-273°C$.
lens	A transparent optical element, usually of glass, that bends light rays. Positive lenses bring parallel rays to a focus; negative lenses cause parallel rays to diverge in a cone.

limb	The edge of the visible disk of a star, moon, or planet.
meniscus	A disk or lens that is concave on one side and convex on the other.
micron	A distance equal to one-millionth of a meter.
nanometer	A distance equal to one-billionth of a meter.
Nasmyth port	A location at the side of an alt-azimuth telescope where light from the primary can be brought to a focus with the aid of a third and fourth mirror.
objective	The largest element in an optical system. In a reflecting telescope, it is the largest mirror.
parabola	A curve obtained by slicing a cone parallel to its side.
paraboloid	A surface created by rotating a parabola about its axis.
period	The interval of time between repetitions of a cycle.
primary	A synonym for *objective*. It is the first optical element a ray of light encounters.
reflector	A telescope with a mirror as the primary optic.
refraction	The bending of a wave as it enters or leaves another medium along its path.
refractor	A telescope with a lens as the primary optic.
resolve	To separate very close objects in the sky.
Ritchey-Chrétien	A type of telescope with hyperbolic primary and secondary mirrors. This design eliminates the coma aberration of parabolic primaries.
secondary	The second optical element (lens or mirror) along the path of a ray.
seeing	An informal term for the sharpness of images as determined by atmospheric turbulence.
spectral line	A characteristic emission of an atom, at a precise wavelength, that corresponds to a jump of an electron between two energy levels in the atom.
spectrometer	An instrument that produces a spectrum and makes a digital record of the variation of brightness with wavelength.
spectrum	A linear display of all the colors (wavelengths) in a sample of light.
wavelength	The distance between two successive maxima of a wave.

INDEX

Credits

Fig. 1.1: Tycho Brahe, *Astronomiae Instauratae Mechanica*, 1598

Fig. 1.4: *Appleton's Cyclopaedia of Applied Mechanics*, 1892

Fig. 1.5: Courtesy of the Research School of Astronomy and Astrophysics, the Australian National University

Fig. 1.6: Courtesy of the Special Collections Research Center, University of Chicago

Fig. 1.7: Courtesy of Alan Whiting

Figs. 2.1–2.2: Courtesy of Gale Gant and the Mount Wilson Observatory Association

Fig. 2.3: *Proceedings of the National Academy of Sciences* 15, no. 3 (March 15, 1929)

Fig. 2.4: Courtesy of Thomas Jarrett and Caltech Astronomy

Fig. 2.5: *Monthly Notices of the Royal Astronomical Society* 113, 658, 1953, reproduced courtesy of Blackwell Publishing, Ltd.

Fig. 3.2: *IAU Symposium* 15, 326, 1962

Fig. 3.4: Courtesy of R. Perley, C. Carilli, J. Dreher, and NRAO/AUI/NSF

Fig. 3.6: G. Clark, *Astrophysical Journal* 153L, 153, 1968

Fig. 3.7: © ESO, European Southern Observatory

Fig. 3.8: V. Rubin and W. K. Ford Jr., *Astrophysical Journal* 159, 379, 1970

Figs. 4.1–4.3: Courtesy of National Optical Astronomy Observatories / AURA / NSF

Fig. 4.4: Courtesy of Cerro Tololo Inter-American Observatory

Fig. 4.5: © ESO, European Southern Observatory

Fig. 4.6: Courtesy of J.-C. Cuillandre

Fig. 4.7: Courtesy of Canada-France-Hawaii Telescope and J.-C. Cuillandre

Figs. 4.8–4.9: © Instituto de Astrofísica de Canarias

Fig. 4.10: Courtesy of the Steward Observatory Mirror Lab, University of Arizona

Fig. 4.11: Courtesy of the Isaac Newton Group of Telescopes, La Palma

Fig. 5.1: Courtesy of Corning Glass Works and the Smithsonian National Air and Space Museum

Fig. 5.2: Courtesy of Lockheed Martin Missiles and Space Company

Fig. 5.3: Courtesy of Ball Aerospace

Fig. 5.4: Courtesy of NASA and the University of Arizona NICMOS team

Fig. 5.5: Courtesy of George Sonneborn and NASA

Fig. 5.6: Courtesy of Alan Fruchter, the Space Telescope Science Institute ERO team, and NASA

Fig. 5.7: Courtesy of R. Williams, the Space Telescope Science Institute Hubble Deep Field team, and NASA

Fig. 5.8: Courtesy of C. Burrows, J. Hester, J. Morse, the Space Telescope Science Institute, and NASA

Fig. 5.9: Courtesy of W. Couch and NASA

Fig. 6.1: Courtesy of William C. Keel, University of Alabama

Fig. 6.2: Courtesy of the California Association for Research in Astronomy

Fig. 6.3: After a diagram in a report from the California Association for Research in Astronomy

Figs. 6.4–6.5: Courtesy of the California Association for Research in Astronomy

Fig. 6.6: Illustration by Tom Connell / Wildlife Art Ltd. from *The Readers Digest Atlas of the Universe,* Weldon Owen Inc., 2000

Fig. 6.7: Courtesy of Richard Wainscoat, Institute for Astronomy, University of Hawaii

Figs. 7.1–7.5: Courtesy of the Steward Observatory Mirror Lab, University of Arizona

Fig. 7.6: Courtesy of the European Industrial Engineering Company, Mestre, Italy

Fig. 7.7: Photos by Sandra Yox, Ray Bertram, J. Waak, and U. Schwarzkopf, courtesy of Steward Observatory Mirror Lab, University of Arizona

Fig. 7.8: Photo by Lori Styles, courtesy of the Steward Observatory Mirror Lab, University of Arizona

Figs. 8.1–8.3: © ESO, European Southern Observatory

Figs. 8.4–8.5: Courtesy of the National Astronomical Observatory of Japan

Fig. 8.6: Courtesy of Corning Glass

Fig. 8.7: Courtesy of the Gemini Observatory / Association of Universities for Research in Astronomy

Fig. 8.8: Courtesy of the Hobby-Eberly Observatory

Fig. 8.9: Courtesy of the Instituto de Astrofísica de Canarias

Figs. 9.1 and 9.3: Courtesy of H. McAllister, Center for High Angular Resolution Astronomy, Georgia State University

Fig. 9.4: Courtesy of the W. M. Keck Observatory and Claire Max, Lawrence Livermore National Laboratory

Fig. 9.5: Courtesy of J. M. Beckers